Lecture Notes in Mathematics 1820

Editors:
J.-M. Morel, Cachan
F. Takens, Groningen
B. Teissier, Paris

T0224618

Springer
Berlin
Heidelberg
New York
Hong Kong
London
Milan
Paris
Tokyo

Fumio Hiai
Hideki Kosaki

Means of
Hilbert Space Operators

 Springer

Authors

Fumio Hiai

Graduate School of Information Sciences
Tohoku University
Aoba-ku, Sendai
980-8579 Japan
e-mail: hiai@math.is.tohoku.ac.jp

Hideki Kosaki

Graduate School of Mathematics
Kyushu University
Higashi-ku, Fukuoka
812-8581 Japan
e-mail: kosaki@math.kyushu-u.ac.jp

Cataloging-in-Publication Data applied for
Bibliographic information published by Die Deutsche Bibliothek

Die Deutsche Bibliothek lists this publication in the Deutsche Nationalbibliografie;
detailed bibliographic data is available in the Internet at http://dnb.ddb.de

Mathematics Subject Classification (2000): 47A30, 47A64, 15A60

ISSN 0075-8434
ISBN 3-540-40680-8 Springer-Verlag Berlin Heidelberg New York

Springer-Verlag Berlin Heidelberg New York a member of BertelsmannSpringer
Science + Business Media GmbH

http://www.springer.de

© Springer-Verlag Berlin Heidelberg 2003
Printed in Germany

Typesetting: Camera-ready TEX output by the author

SPIN: 10949634   41/3142/ du - 543210 - Printed on acid-free paper

Preface

Roughly speaking two kinds of operator and/or matrix inequalities are known, of course with many important exceptions. Operators admit several natural notions of orders (such as positive semidefiniteness order, some majorization orders and so on) due to their non-commutativity, and some operator inequalities clarify these order relations. There is also another kind of operator inequalities comparing or estimating various quantities (such as norms, traces, determinants and so on) naturally attached to operators.

Both kinds are of fundamental importance in many branches of mathematical analysis, but are also sometimes highly non-trivial because of the non-commutativity of the operators involved. This monograph is mainly devoted to means of Hilbert space operators and their general properties with the main emphasis on their norm comparison results. Therefore, our operator inequalities here are basically of the second kind. However, they are not free from the first in the sense that our general theory on means relies heavily on a certain order for operators (i.e., a majorization technique which is relevant for dealing with unitarily invariant norms).

In recent years many norm inequalities on operator means have been investigated. We develop here a general theory which enables us to treat them in a unified and axiomatic fashion. More precisely, we associate operator means to given scalar means by making use of the theory of Stieltjes double integral transformations. Here, Peller's characterization of Schur multipliers plays an important role, and indeed guarantees that our operator means are bounded operators. Basic properties on these operator means (such as the convergence property and norm bounds) are studied. We also obtain a handy criterion (in terms of the Fourier transformation) to check the validity of norm comparison among operator means.

Sendai, June 2003 *Fumio Hiai*

Fukuoka, June 2003 *Hideki Kosaki*

Contents

1

Introduction

The present monograph is devoted to a thorough study of means for Hilbert space operators, especially comparison of (unitarily invariant) norms of operator means and their convergence properties in various aspects.

The Hadamard product (or Schur product) $A \circ B$ of two matrices $A = [a_{ij}]$, $B = [b_{ij}]$ means their entry-wise product $[a_{ij}b_{ij}]$. This notion is a common and powerful technique in investigation of general matrix (and/or operator) norm inequalities, and particularly so in that of perturbation inequalities and commutator estimates. Assume that $n \times n$ matrices $H, K, X \in M_n(\mathbf{C})$ are given with $H, K \geq 0$ and diagonalizations

$$H = U\,\mathrm{diag}(s_1, s_2, \ldots, s_n)U^* \text{ and } K = V\,\mathrm{diag}(t_1, t_2, \ldots, t_n)V^*.$$

In our previous work [39], to a given scalar mean $M(s, t)$ (for $s, t \in \mathbf{R}_+$), we associated the corresponding matrix mean $M(H, K)X$ by

$$M(H, K)X = U\left([M(s_i, t_j)] \circ (U^*XV)\right)V^*. \tag{1.1}$$

For a scalar mean $M(s, t)$ of the form $\sum_{i=1}^n f_i(s)g_i(t)$ one easily observes $M(H, K)X = \sum_{i=1}^n f_i(H)Xg_i(K)$, and we note that this expression makes a perfect sense even for Hilbert space operators H, K, X with $H, K \geq 0$. However, for the definition of general matrix means $M(H, K)X$ (such as A-L-G interpolation means $M_\alpha(H, K)X$ and binomial means $B_\alpha(H, K)X$ to be explained later) the use of Hadamard products or something alike seems unavoidable.

The first main purpose of the present monograph is to develop a reasonable theory of means for Hilbert space operators, which works equally well for general scalar means (including M_α, B_α and so on). Here two difficulties have to be resolved: (i) Given (infinite-dimensional) diagonal operators $H, K \geq 0$, the definition (1.1) remains legitimate for $X \in \mathcal{C}_2(\mathcal{H})$, the Hilbert-Schmidt class operators on a Hilbert space $\mathcal{H}$, as long as entries $M(s_i, t_j)$ stay bounded (and $M(H, K)X \in \mathcal{C}_2(\mathcal{H})$). However, what we want is a mean $M(H, K)X$ ($\in B(\mathcal{H})$) for each bounded operator $X \in B(\mathcal{H})$. (ii) General positive operators H, K are no longer diagonal so that continuous spectral decomposition has to be used. The requirement in (i) says that the concept of a Schur multiplier ([31, 32, 66]) has to enter our picture, and hence what we need is a continuous analogue of the operation (1.1) with this concept built in. The theory of (Stieltjes) double integral transformations ([14]) due to M. Sh. Birman, M. Z. Solomyak and others is suited for this purpose. With this apparatus the operator mean $M(H, K)X$ is defined (in Chapter 3) as

$$M(H, K)X = \int_0^{\|H\|} \int_0^{\|K\|} M(s, t) \, dE_s X \, dF_t \qquad (1.2)$$

with the spectral decompositions

$$H = \int_0^{\|H\|} s \, dE_s \quad \text{and} \quad K = \int_0^{\|K\|} t \, dF_t.$$

Double integral transformations as above were actually considered with general functions $M(s, t)$ (which are not necessarily means). This subject has important applications to theories of perturbation, Volterra operators, Hankel operators and so on (see §2.5 for more information including references), and one of central problems here (besides the justification of the double integral (1.2)) is to determine for which unitarily invariant norm the transformation $X \mapsto M(H, K)X$ is bounded. Extensive study has been made in this direction, and V. V. Peller's work ([69, 70]) deserves special mentioning. Namely, he completely characterized ($\mathcal{C}_1$-)Schur multipliers in this setting (i.e., boundedness criterion relative to the trace norm $\|\cdot\|_1$, or equivalently, the operator norm $\|\cdot\|$ by the duality), which is a continuous counterpart of U. Haagerup's characterization ([31, 32]) in the matrix setting. Our theory of operator means is built upon V. V. Peller's characterization (Theorem 2.2) although just an easy part is needed. Unfortunately, his work [69] with a proof (while [70] is an announcement) was not widely circulated, and details of some parts were omitted. Moreover, quite a few references there are not easily accessible. For these reasons and to make the monograph as self-contained as possible, we present details of his proof in Chapter 2 (see §2.1).

As emphasized above, the notions of Hadamard products and double integral transformations play important roles in perturbation theory and commutator estimates. In this monograph we restrict ourselves mainly to symmetric homogeneous means (except in Chapter 8 and §A.1) so that these important topics will not be touched. However, most of the arguments in Chapters 2 and 3 are quite general and our technique can be applicable to these topics (which will be actually carried out in our forthcoming article [55]). It is needless to say that there are large numbers of literature on matrix and/or operator norm inequalities (not necessarily of perturbation and/or commutator-type) based on closely related techniques. We also remark that the technique here is useful for dealing with certain operator equations such as Lyapunov-type equations (see §3.7 and [39, §4]). These related topics as well as relationship to other standard methods for study of operator inequalities (such as majorization theory and so on) are summarized at the end of each chapter together with suitable references, which might be of some help to the reader.

In the rest we will explain historical background at first and then more details on the contents of the present monograph. In the classical work [36] E. Heinz showed the (operator) norm inequality

$$\|H^\theta X K^{1-\theta} + H^{1-\theta} X K^\theta\| \le \|HX + XK\| \quad \text{(for } \theta \in [0,1]) \tag{1.3}$$

for positive operators $H, K \ge 0$ and an arbitrary operator X on a Hilbert space. In the 1979 article [64] A. McIntosh presented a simple proof of

$$\|H^* X K\| \le \frac{1}{2}\|HH^* X + XKK^*\|,$$

which is obviously equivalent to the following estimate for positive operators:

$$\|H^{1/2} X K^{1/2}\| \le \frac{1}{2}\|HX + XK\| \quad (H, K \ge 0).$$

It is the special case $\theta = 1/2$ of (1.3), and he pointed out that a simple and unified approach to so-called Heinz-type inequalities such as (1.3) (and the "difference version" (8.7)) is possible based on this arithmetic-geometric mean inequality. The closely related eigenvalue estimate

$$\mu_n(H^{1/2} K^{1/2}) \le \frac{1}{2}\mu_n(H + K) \quad (n = 1, 2, \ldots)$$

for positive matrices is known ([12]). Here, $\{\mu_n(\cdot)\}_{n=1,2,\ldots}$ denotes singular numbers, i.e., $\mu_n(Y)$ is the n-th largest eigenvalue (with multiplicities counted) of the positive part $|Y| = (Y^* Y)^{1/2}$. This means $|H^{1/2} K^{1/2}| \le \frac{1}{2} U(H+K)U^*$ for some unitary matrix U so that we have

$$|||H^{1/2} K^{1/2}||| \le \frac{1}{2}|||H + K|||$$

for an arbitrary unitarily invariant norm $||| \cdot |||$.

In the 1993 article [10] R. Bhatia and C. Davis showed the following strengthening:

$$|||H^{1/2} X K^{1/2}||| \le \frac{1}{2}|||HX + XK||| \tag{1.4}$$

for matrices, which of course remains valid for Hilbert space operators $H, K \ge 0$ and X by the standard approximation argument. On the other hand, in [3] T. Ando obtained the matrix Young inequality

$$\mu_n\left(H^{\frac{1}{p}} K^{\frac{1}{q}}\right) \le \mu_n\left(\frac{1}{p}H + \frac{1}{q}K\right) \quad (n = 1, 2, \ldots) \tag{1.5}$$

for $p, q > 1$ with $p^{-1} + q^{-1} = 1$. Although the weak matrix Young inequality

$$|||H^{\frac{1}{p}}XK^{\frac{1}{q}}||| \leq \kappa_p|||\tfrac{1}{p}HX + \tfrac{1}{q}XK||| \tag{1.6}$$

holds with some constant $\kappa_p \geq 1$ ([54]), without this constant the inequality fails to hold for the operator norm $||| \cdot ||| = \| \cdot \|$ (unless $p = 2$) as was pointed out in [2]. Instead, the following slightly weaker inequality holds always:

$$|||H^{\frac{1}{p}}XK^{\frac{1}{q}}||| \leq \tfrac{1}{p}|||HX||| + \tfrac{1}{q}|||XK|||. \tag{1.7}$$

In the recent years the above-mentioned arithmetic-geometric mean and related inequalities have been under active investigation by several authors, and very readable accounts on this subject can be found in [2, 8, 84]. Motivated by these works, in a series of recent articles [54, 38, 39] we have investigated simple unified proofs for known (as well as many new) norm inequalities in a similar nature, and our investigation is summarized in the recent survey article [40]. We also point out that closely related analysis was made in the recent article [13] by R. Bhatia and K. Parthasarathy. For example as a refinement of (1.4) the arithmetic-logarithmic-geometric mean inequality

$$|||H^{1/2}XK^{1/2}||| \leq ||| \int_0^1 H^x XK^{1-x}dx||| \leq \frac{1}{2}|||HX + XK||| \tag{1.8}$$

was obtained in [38]. The technique in this article actually permitted us to compare these quantities with

$$|||\frac{1}{m}\sum_{k=1}^{m} H^{\frac{k}{m+1}}XK^{\frac{m+1-k}{m+1}}|||, \quad |||\frac{1}{n}\sum_{k=0}^{n-1} H^{\frac{k}{n-1}}XK^{\frac{n-1-k}{n-1}}|||, \tag{1.9}$$

and moreover in the appendix to [38] we discussed the $||| \cdot |||$-convergence

$$\begin{cases} \dfrac{1}{m}\sum_{k=1}^{m} H^{\frac{k}{m+1}}XK^{\frac{m+1-k}{m+1}} \longrightarrow \displaystyle\int_0^1 H^x XK^{1-x}dx \quad (\text{as } m \to \infty), \\[4mm] \dfrac{1}{n}\sum_{k=0}^{n-1} H^{\frac{k}{n-1}}XK^{\frac{n-1-k}{n-1}} \longrightarrow \displaystyle\int_0^1 H^x XK^{1-x}dx \quad (\text{as } n \to \infty) \end{cases} \tag{1.10}$$

under certain circumstances.

The starting point of the analysis made in [39] was an axiomatic treatment on matrix means (i.e., matrix means $M(H,K)X$ (see (1.1)) associated to scalar means $M(s,t)$ satisfying certain axioms), and a variety of generalizations of the norm inequalities explained so far were obtained as applications. As in [39] a certain class of symmetric homogeneous (scalar) means is considered in the present monograph, but our main concern here is a study of corresponding means for Hilbert space operators instead. In order to be able to define $M(H,K)X$ ($\in B(\mathcal{H})$) for each $X \in B(\mathcal{H})$ (by the double integral transformation (1.2)), our mean $M(s,t)$ has to be a Schur multiplier in addition. For two such means $M(s,t), N(s,t)$ we introduce the partial order:

$M \preceq N$ if and only if $M(e^x, 1)/N(e^x, 1)$ is positive definite. If this is the case, then for non-singular positive operators H, K we have the integral expression

$$M(H, K)X = \int_{-\infty}^{\infty} H^{ix}(N(H, K)X)K^{-ix}d\nu(x) \qquad (1.11)$$

with a probability measure ν (see Theorems 3.4 and 3.7 for the precise statement), and of course the Bochner theorem is behind. Under such circumstances (thanks to the general fact explained in §A.2) we actually have

$$|||M(H, K)X||| \leq |||N(H, K)X||| \qquad (1.12)$$

(even without the non-singularity of $H, K \geq 0$). This inequality actually characterizes the order $M \preceq N$, and is a source for a variety of concrete norm inequalities (as was demonstrated in [40]). The order $\preceq$ and (1.11), (1.12) were also used in [39] for matrices, but much more involved arguments are required for Hilbert space operators, which will be carried out in Chapter 3. It is sometimes not an easy task to determine if a given mean $M(s, t)$ is a Schur multiplier. However, the mean $M_\infty(s, t) = \max\{s, t\}$ comes to the rescue: (i) The mean M_∞ itself is a Schur multiplier. (ii) A mean majorized by M_∞ (relative to $\preceq$) is a Schur multiplier. These are consequences of (1.11), (1.12), and enable us to prove that all the means considered in [39] are indeed Schur multipliers. The observation (i) also follows from the discrete decomposition of $\max\{s, t\}$ worked out in §A.3, which might be of independent interest. Furthermore, a general norm estimate of the transformation $X \mapsto M(H, K)X$ is established for means $M \preceq M_\infty$. In Chapter 4 we study the convergence $M(H_n, K_n)X \to M(H, K)X$ (in $||| \cdot |||$ or in the strong operator topology) under the strong convergence $H_n \to H$, $K_n \to K$ of the positive operators involved.

The requirement for the convergence (1.10) in the appendix to [39] was the following finiteness condition: either $|||H|||, |||K||| < \infty$ or $|||X||| < \infty$. This requirement is somewhat artificial (and too restrictive), and the arguments presented there were ad hoc. The second main purpose of the monograph is to present systematic and thorough investigation on such convergence phenomena. In [39] we dealt with the following one-parameter families of scalar means:

$$M_\alpha(s, t) = \frac{\alpha - 1}{\alpha} \times \frac{s^\alpha - t^\alpha}{s^{\alpha-1} - t^{\alpha-1}} \quad (-\infty \leq \alpha \leq \infty),$$

$$A_\alpha(s, t) = \frac{1}{2}(s^\alpha t^{1-\alpha} + s^{1-\alpha}t^\alpha) \quad (0 \leq \alpha \leq 1),$$

$$B_\alpha(s, t) = \left(\frac{s^\alpha + t^\alpha}{2}\right)^{1/\alpha} \quad (-\infty \leq \alpha \leq \infty).$$

It is straight-forward to see that $M_\alpha(s, t), A_\alpha(s, t)$ are Schur multipliers, and also so is $B_{1/n}(s, t)$ thanks to the the binomial expansion $B_{1/n}(s, t) =$

$2^{-n}\sum_{k=0}^{n}\binom{n}{k}s^{\frac{k}{n}}t^{\frac{n-k}{n}}$. We indeed show that all of $B_\alpha(s,t)$ are (by proving $B_\alpha \preceq M_\infty$). Thus, all of the above give rise to operator means. Note $M_{1/2}(s,t) = \sqrt{st}$ (the geometric mean), $M_2 = \frac{1}{2}(s+t)$ (the arithmetic mean) and

$$M_1(s,t) \ \left(= \lim_{\alpha \to 1} M_\alpha(s,t)\right)$$

$$= \frac{s-t}{\log s - \log t} = \int_0^1 s^x t^{1-x} dx \quad \text{(the logarithmic mean)}.$$

Because of these reasons $\{M_\alpha(s,t)\}_{-\infty \le \alpha \le \infty}$ will be referred to as the A-L-G interpolation means. The convergence (1.10) (see also (5.1)) means

$$\lim_{m \to \infty} |||M_{\frac{m}{m+1}}(H,K)X - L||| = \lim_{n \to \infty} |||M_{\frac{n}{n-1}}(H,K)X - L||| = 0$$

with the logarithmic mean $L = M_1(H,K)X = \int_0^1 H^x X K^{1-x} dx$, and the main result in Chapter 5 is the following generalization:

$$\lim_{\alpha \to \alpha_0} |||M_\alpha(H,K)X - M_{\alpha_0}(H,K)X||| = 0$$

under the assumption $|||M_\beta(H,K)X||| < \infty$ for some $\beta > \alpha_0$. This is a "dominated convergence theorem" for the A-L-G means, the proof of which is indeed based on Lebesgue's theorem applied to the relevant integral expression (1.11) with the concrete form of the density $d\nu(x)/dx$. Similar dominated convergence theorems for the Heinz-type means $A_\alpha(H,K)X = \frac{1}{2}(H^\alpha X K^{1-\alpha} + H^{1-\alpha} X K^\alpha)$ (or rather the single components $H^\alpha X K^{1-\alpha}$) and the binomial means $B_\alpha(H,K)X$ are also obtained together with other related results in Chapters 6 and 7.

A slightly different subject is covered in Chapter 8, that might be of independent interest. The homogeneous alternating sums

$$\begin{cases} \mathbf{A}(n) = \sum_{k=1}^{n}(-1)^{k-1} H^{\frac{k}{n+1}} X K^{\frac{n+1-k}{n+1}} & \text{(with } n = 1,2,\cdots), \\[2em] \mathbf{B}(m) = \sum_{k=0}^{m-1}(-1)^{k} H^{\frac{k}{m-1}} X K^{\frac{m-1-k}{m-1}} & \text{(with } m = 2,3,\cdots) \end{cases}$$

are not necessarily symmetric (depending upon parities of n, m), but our method works and integral expressions akin to (1.11) (sometimes with signed measures ν) are available. This enables us to determine behavior of unitarily invariant norms of these alternating sums of operators such as mutual comparison, uniform bounds, monotonicity and so on.

Some technical results used in the monograph are collected in Appendices, and §A.1 is concerned with extension of our arguments to certain nonsymmetric means.

Double integral transformations

Throughout the monograph a Hilbert space $\mathcal{H}$ is assumed to be separable. The algebra $B(\mathcal{H})$ of all bounded operators on $\mathcal{H}$ is a Banach space with the operator norm $\|\cdot\|$. For $1 \leq p < \infty$ let $\mathcal{C}_p(\mathcal{H})$ denote the Schatten p-class consisting of (compact) operators $X \in B(\mathcal{H})$ satisfying $\mathrm{Tr}(|X|^p) < \infty$ with $|X| = (X^*X)^{1/2}$, where Tr is the usual trace. The space $\mathcal{C}_p(\mathcal{H})$ is an ideal of $B(\mathcal{H})$ and a Banach space with the Schatten p-norm $\|X\|_p = (\mathrm{Tr}(|X|^p))^{1/p}$. In particular, $\mathcal{C}_1(\mathcal{H})$ is the trace class, and $\mathcal{C}_2(\mathcal{H})$ is the Hilbert-Schmidt class which is a Hilbert space with the inner product $(X,Y)_{\mathcal{C}_2(\mathcal{H})} = \mathrm{Tr}(XY^*)$ $(X,Y \in \mathcal{C}_2(\mathcal{H}))$. The algebra $B(\mathcal{H})$ is faithfully (hence isometrically) represented on the Hilbert space $\mathcal{C}_2(\mathcal{H})$ by the left (also right) multiplication: $X \in \mathcal{C}_2(\mathcal{H}) \mapsto AX, XA \in \mathcal{C}_2(\mathcal{H})$ for $A \in B(\mathcal{H})$. Standard references on these basic topics (as well as unitarily invariant norms) are [29, 37, 77].

In this chapter we choose and fix positive operators H, K on $\mathcal{H}$ with the spectral decompositions

$$H = \int_0^{\|H\|} s\, dE_s \quad \text{and} \quad K = \int_0^{\|K\|} t\, dF_t$$

respectively. We will use both of the notations dE_s, E_Λ (for Borel sets $\Lambda \subseteq [0, \|H\|]$) interchangeably in what follows (and do the same for the other spectral measure F). Let λ (resp. μ) be a finite positive measure on the interval $[0, \|H\|]$ (resp. $[0, \|K\|]$) equivalent (in the absolute continuity sense) to dE_s (resp. dF_t). For instance the measures

$$\lambda(\Lambda) = \sum_{n=1}^{\infty} \frac{1}{n^2}(E_\Lambda e_n, e_n) \quad (\Lambda \subseteq [0, \|H\|]),$$

$$\mu(\Xi) = \sum_{n=1}^{\infty} \frac{1}{n^2}(F_\Xi e_n, e_n) \quad (\Xi \subseteq [0, \|K\|])$$

do the job, where $\{e_n\}_{n=1,2,\cdots}$ is an orthonormal basis for $\mathcal{H}$. We choose and fix a function $\phi(s,t)$ in $L^\infty([0, \|H\|] \times [0, \|K\|]; \lambda \times \mu)$. For each operator

$X \in B(\mathcal{H})$, the algebra of all bounded operators on $\mathcal{H}$, we would like to justify its "double integral" transformation formally written as

$$\Phi(X) = \int_0^{\|H\|} \int_0^{\|K\|} \phi(s,t) \, dE_s X dF_t$$

(see [14]). As long as $X \in \mathcal{C}_2(\mathcal{H})$, the Hilbert-Schmidt class operators, desired justification is quite straight-forward and moreover under such circumstances we have $\Phi(X) \in \mathcal{C}_2(\mathcal{H})$ with the norm bound

$$\|\Phi(X)\|_2 \leq \|\phi\|_{L^\infty(\lambda \times \mu)} \times \|X\|_2. \tag{2.1}$$

In fact, with the left multiplication π_ℓ and the right multiplication π_r, $\pi_\ell(E_\Lambda)$ and $\pi_r(F_\Xi)$ (with Borel sets $\Lambda \subseteq [0, \|H\|]$ and $\Xi \subseteq [0, \|K\|]$) are commuting projections acting on the Hilbert space $\mathcal{C}_2(\mathcal{H})$ so that $\pi_\ell(E_\Lambda)\pi_r(F_\Xi)$ is a projection. It is plain to see that one gets a spectral family acting on the Hilbert space $\mathcal{C}_2(\mathcal{H})$ from those "rectangular" projections so that the ordinary functional calculus via $\phi(s,t)$ gives us a bounded linear operator on $\mathcal{C}_2(\mathcal{H})$. With this interpretation we set

$$\Phi(X) = \left(\int_0^{\|H\|} \int_0^{\|K\|} \phi(s,t) \, d(\pi_\ell(E)\pi_r(F)) \right) X. \tag{2.2}$$

Note that the Hilbert-Schmidt class operator X in the right side here is regarded as a vector in the Hilbert space $\mathcal{C}_2(\mathcal{H})$, and (2.1) is obvious.

In applications of double integral transformations (for instance to stability problems of perturbation) it is important to be able to specify classes of functions ϕ for which the domain of $\Phi(\cdot)$ can be enlarged to various operator ideals (such as $\mathcal{C}_p$-ideals). In fact, some useful sufficient conditions (in terms of certain Lipschitz conditions on $\phi(\cdot, \cdot)$) were announced in [14] (whose proofs were sketched in [15]), but unfortunately they are not so helpful for our later purpose. More detailed information on double integral transformations will be given in §2.5.

2.1 Schur multipliers and Peller's theorem

We begin with the definition of Schur multipliers (acting on operators on $\mathcal{H}$).

Definition 2.1. When $\Phi \, (= \Phi|_{\mathcal{C}_1(\mathcal{H})}) : X \mapsto \Phi(X)$ gives rise to a bounded transformation on the ideal $\mathcal{C}_1(\mathcal{H})$ ($\subseteq \mathcal{C}_2(\mathcal{H})$) of trace class operators, $\phi(s,t)$ is called a *Schur multiplier* (relative to the pair (H, K)).

When this requirement is met, by the usual duality $B(\mathcal{H}) = (\mathcal{C}_1(\mathcal{H}))^*$ the transpose of Φ gives rise to a bounded transformation on $B(\mathcal{H})$ (i.e., the largest possible domain) as will be explained in the next §2.2. The next important characterization due to V. V. Peller will play a fundamental role in our investigation on means of operators:

Theorem 2.2. (V.V. Peller, [69, 70]) *For $\phi \in L^\infty([0, \|H\|] \times [0, \|K\|]; \lambda \times \mu)$ the following conditions are all equivalent:*

(i)   *ϕ is a Schur multiplier;*

(ii) *whenever a measurable function $k : [0, \|H\|] \times [0, \|K\|] \to \mathbf{C}$ is the kernel of a trace class operator $L^2([0, \|H\|]; \lambda) \to L^2([0, \|K\|]; \mu)$, so is the product $\phi(s, t)k(s, t)$;*

(iii) *one can find a finite measure space (Ω, σ) and functions $\alpha \in L^\infty([0, \|H\|] \times \Omega; \lambda \times \sigma)$, $\beta \in L^\infty([0, \|K\|] \times \Omega; \mu \times \sigma)$ such that*

$$\phi(s, t) = \int_\Omega \alpha(s, x)\beta(t, x)d\sigma(x) \quad \text{for all } s \in [0, \|H\|], t \in [0, \|K\|]; \quad (2.3)$$

(iv) *one can find a measure space (Ω, σ) and measurable functions α, β on $[0, \|H\|] \times \Omega$, $[0, \|K\|] \times \Omega$ respectively such that the above (2.3) holds and*

$$\left\| \int_\Omega |\alpha(\cdot, x)|^2 d\sigma(x) \right\|_{L^\infty(\lambda)} \left\| \int_\Omega |\beta(\cdot, x)|^2 d\sigma(x) \right\|_{L^\infty(\mu)} < \infty.$$

A few remarks are in order. (a) The implication (iii) $\Rightarrow$ (iv) is trivial. (b) The finiteness condition in (iv) and the Cauchy-Schwarz inequality guarantee the integrability of the integrand in the right-hand side of (2.3). (c) The condition (iii) is stronger than what was stated in [69, 70], but the proof in [69] (presented below) actually says (ii) $\Rightarrow$ (iii).

Unfortunately Peller's article [69] (with a proof) was not widely circulated. Because of this reason and partly to make the present monograph as much as self-contained, the proof of the theorem is presented in what follows.

Proof of (iv) $\Rightarrow$ (i)

Although this is a relatively easy part in the proof, we present detailed arguments here because its understanding will be indispensable for our later arguments. So let us assume that $\phi(s, t)$ admits an integral representation stated in (iv). For a rank-one operator $X = \xi \otimes \eta^c$ we have $\pi_\ell(E_\Lambda)\pi_r(F_\Xi)X = (E_\Lambda\xi) \otimes (F_\Xi\eta)^c$ so that from (2.3) we get

$$\Phi(X) = \int_0^{\|H\|} \int_0^{\|K\|} \int_\Omega \alpha(s, x)\beta(t, x) (dE_s\xi) \otimes (dF_t\eta)^c \, d\sigma(x)$$

$$= \int_\Omega \xi(x) \otimes \eta(x)^c \, d\sigma(x)$$

with

$$\xi(x) = \int_0^{\|H\|} \alpha(s, x) \, dE_s\xi \quad \text{and} \quad \eta(x) = \int_0^{\|K\|} \overline{\beta(t, x)} \, dF_t\eta. \quad (2.4)$$

More precisely, the above integral can be understood for example in the weak sense:

$$(\Phi(X)\xi',\eta') = \int_\Omega ((\xi(x)\otimes\eta(x)^c)\xi',\eta')\,d\sigma(x)$$

$$= \int_\Omega (\xi',\eta(x))(\xi(x),\eta')\,d\sigma(x). \tag{2.5}$$

The above $\xi(x), \eta(x)$ are vectors for a.e. $x \in \Omega$ as will be seen shortly. We use Theorem A.5 in §A.2 and the Cauchy-Schwarz inequality to get

$$\|\Phi(\xi\otimes\eta^c)\|_1 \le \int_\Omega \|\xi(x)\otimes\eta(x)^c\|_1 d\sigma(x) = \int_\Omega \|\xi(x)\|\times\|\eta(x)\|\,d\sigma(x)$$

$$\le \left(\int_\Omega \|\xi(x)\|^2 d\sigma(x)\right)^{1/2}\left(\int_\Omega \|\eta(x)\|^2 d\sigma(x)\right)^{1/2}. \tag{2.6}$$

Since $\|\xi(x)\|^2 = \int_0^{\|H\|} |\alpha(s,x)|^2 d(E_s\xi,\xi)$ with the total mass of $d(E_s\xi,\xi)$ being $\|\xi\|^2$, we have

$$\int_\Omega \|\xi(x)\|^2 d\sigma(x) = \int_0^{\|H\|}\left(\int_\Omega |\alpha(s,x)|^2 d\sigma(x)\right) d(E_s\xi,\xi)$$

$$\le \left\|\int_\Omega |\alpha(\cdot,x)|^2 d\sigma(x)\right\|_{L^\infty(\lambda)}\times\|\xi\|^2 \tag{2.7}$$

by the Fubini-Tonneli theorem. A similar bound for $\int_\Omega \|\eta(x)\|^2 d\sigma(x)$ is also available, and consequently from (2.6), (2.7) we get

$$\|\Phi(\xi\otimes\eta^c)\|_1 \le \|\xi\|\times\|\eta\|\times\left\|\int_\Omega |\alpha(\cdot,x)|^2 d\sigma(x)\right\|_{L^\infty(\lambda)}^{1/2}$$

$$\times\left\|\int_\Omega |\beta(\cdot,x)|^2 d\sigma(x)\right\|_{L^\infty(\mu)}^{1/2}.$$

Therefore, we have shown

$$\|\Phi(X)\|_1 \le \left\|\int_\Omega |\alpha(\cdot,x)|^2 d\sigma(x)\right\|_{L^\infty(\lambda)}^{1/2}$$

$$\times\left\|\int_\Omega |\beta(\cdot,x)|^2 d\sigma(x)\right\|_{L^\infty(\mu)}^{1/2}\times\|X\|_1 \tag{2.8}$$

for rank-one operators X. Note that (2.7) (together with the finiteness requirement in the theorem) shows $\|\xi(x)\| < \infty$, i.e., $\xi(x)$ is indeed a vector for a.e. $x \in \Omega$. Also (2.8) guarantees that $\Phi(X) = \int_\Omega \xi(x)\otimes\eta(x)^c d\sigma(x)$ falls into the ideal $\mathcal{C}_1(\mathcal{H})$ of trace class operators.

We claim that the estimate (2.8) remains valid for finite-rank operators. Indeed, thanks to the standard polar decomposition and diagonalization technique, such an operator X admits a representation $X = \sum_{i=1}^n \xi_i\otimes\eta_i^c$ satisfying $\|X\|_1 = \sum_{i=1}^n \|\xi_i\|\times\|\eta_i\|$. Then, we estimate

$$\|\Phi(X)\|_1 \le \sum_{i=1}^{n} \|\Phi(\xi_i \otimes \eta_i^c)\|_1$$

$$\le \sum_{i=1}^{n} \|\xi_i\| \times \|\eta_i\| \times \left\| \int_\Omega |\alpha(\cdot,x)|^2 d\sigma(x) \right\|_{L^\infty(\lambda)}^{1/2}$$

$$\times \left\| \int_\Omega |\beta(\cdot,x)|^2 d\sigma(x) \right\|_{L^\infty(\mu)}^{1/2}$$

(by (2.8) for rank-one operators)

$$= \left\| \int_\Omega |\alpha(\cdot,x)|^2 d\sigma(x) \right\|_{L^\infty(\lambda)}^{1/2} \times \left\| \int_\Omega |\beta(\cdot,x)|^2 d\sigma(x) \right\|_{L^\infty(\mu)}^{1/2} \times \|X\|_1.$$

We now assume $X \in \mathcal{C}_1(\mathcal{H})$. Choose a sequence $\{X_n\}_{n=1,2,\dots}$ of finite-rank operators converging to X in $\|\cdot\|_1$. Since convergence also takes place in $\|\cdot\|_2$ ($\le \|\cdot\|_1$), we see that $\Phi(X_n)$ tends to $\Phi(X)$ in $\|\cdot\|_2$ (by (2.1)) and consequently in the operator norm $\|\cdot\|$. The lower semi-continuity of $\|\cdot\|_1$ relative to the $\|\cdot\|$-topology thus yields

$$\|\Phi(X)\|_1 \le \liminf_{n\to\infty} \|\Phi(X_n)\|_1$$

$$\le \liminf_{n\to\infty} \left(\left\| \int_\Omega |\alpha(\cdot,x)|^2 d\sigma(x) \right\|_{L^\infty(\lambda)}^{1/2} \right.$$

$$\left. \times \left\| \int_\Omega |\beta(\cdot,x)|^2 d\sigma(x) \right\|_{L^\infty(\mu)}^{1/2} \times \|X_n\|_1 \right)$$

(by (2.8) for finite-rank operators)

$$= \left(\left\| \int_\Omega |\alpha(\cdot,x)|^2 d\sigma(x) \right\|_{L^\infty(\lambda)}^{1/2} \times \left\| \int_\Omega |\beta(\cdot,x)|^2 d\sigma(x) \right\|_{L^\infty(\mu)}^{1/2} \right) \|X\|_1.$$

Therefore, $\Phi(X)$ belongs to $\mathcal{C}_1(\mathcal{H})$, and moreover $\Phi(\cdot)$ restricted to $\mathcal{C}_1(\mathcal{H})$ gives rise to a bounded transformation as desired.

Proof of (i) $\Rightarrow$ (ii)
One can choose a sequence $\{\xi_m\}$ in $\mathcal{H}$ with $\sum_m \|\xi_m\|^2 < \infty$ such that $\{E_\Lambda \xi_m : \Lambda \subseteq [0, \|H\|]\}$ ($m = 1, 2, \dots$) are mutually orthogonal and λ is equivalent to the measure $\sum_m (E_\Lambda \xi_m, \xi_m)$. In fact, choose a sequence $\{\xi_m\}$ for which $\sum_m \|\xi_m\|^2 < \infty$ and $\sum_m (E_\Lambda \xi_m, \xi_m)$ is equivalent to λ. We set

$$\Lambda_m = \left\{ s \in [0, \|H\|] : \frac{d(E_s \xi_m, \xi_m)}{d\lambda(s)} > 0 \right\}$$

with the Radon-Nikodym derivative $d(E_s \xi_m, \xi_m)/d\lambda(s)$ with respect to λ. Choose mutually disjoint measurable subsets $\Lambda_m^0 \subseteq \Lambda_m$ ($m = 1, 2, \dots$) with $\bigcup_m \Lambda_m^0 = \bigcup_m \Lambda_m$; then a required sequence is obtained by replacing ξ_m by $E_{\Lambda_m^0} \xi_m$. Furthermore, we easily observe that the condition (ii) (as well as (i)) is unchanged for equivalent measures (by considering the unitary multiplication operator induced by the square root of the relevant Radon-Nikodym derivative). So one can assume $\lambda(\Lambda) = \sum_m (E_\Lambda \xi_m, \xi_m)$ with $\{\xi_m\}$ as above and similarly $\mu(\Xi) = \sum_n (F_\Xi \eta_n \eta_n)$ where $\sum_m \|\eta_n\|^2 < \infty$ and $\{F_\Xi \eta_n : \Xi \subseteq [0, \|K\|]\}$ ($n = 1, 2, \dots$) are mutually orthogonal.

Let $\mathcal{H}_1$ be the closed subspace of $\mathcal{H}$ spanned by $\{E_\Lambda \xi_m : \Lambda \subseteq [0, \|H\|], m \geq 1\}$ and $\mathcal{H}_2$ be spanned by $\{F_\Xi \eta_n : \Xi \subseteq [0, \|K\|], n \geq 1\}$; then $L^2(\lambda) = L^2([0, \|H\|]; \lambda)$ and $L^2(\mu) = L^2([0, \|K\|]; \mu)$ are isometrically isomorphic to $\mathcal{H}_1$ and $\mathcal{H}_2$ respectively by the correspondences

$$\chi_\Lambda \leftrightarrow \sum_m E_\Lambda \xi_m \quad \text{and} \quad \chi_\Xi \leftrightarrow \sum_n F_\Xi \eta_n.$$

Assume that a measurable function k on $[0, \|H\|] \times [0, \|K\|]$ is the kernel of a trace class operator $R : L^2(\lambda) \to L^2(\mu)$, i.e.,

$$(Rf)(t) = \int_0^{\|H\|} k(s,t) f(s) \, d\lambda(s) \quad \text{for } f \in L^2(\lambda).$$

The assumption implies in particular that $k(s,t)$ and hence $\phi(s,t)k(s,t)$ are square integrable with respect to $\lambda \times \mu$ so that the latter is the kernel of a Hilbert-Schmidt class operator. We prove under the assumption (i) that $\phi(s,t)k(s,t)$ is indeed the kernel of a trace class operator. Define $X \in \mathcal{C}_1(\mathcal{H})$ by composing R with the orthogonal projection $P_{\mathcal{H}_1}$ as follows:

$$\mathcal{H} \xrightarrow{P_{\mathcal{H}_1}} \mathcal{H}_1 \cong L^2(\lambda) \xrightarrow{R} L^2(\mu) \cong \mathcal{H}_2 \hookrightarrow \mathcal{H}.$$

Then (i) yields $\Phi(X) \in \mathcal{C}_1(\mathcal{H})$. For each $\Lambda \subseteq [0, \|H\|]$ and $\Xi \subseteq [0, \|K\|]$ we have

$$\left(\Phi(X) \left(\sum_m E_\Lambda \xi_m \right), \sum_n F_\Xi \eta_n \right)$$

$$= \sum_{m,n} \left(\Phi(X), (E_\Lambda \xi_m) \otimes (F_\Xi \eta_n)^c \right)_{\mathcal{C}_2(\mathcal{H})}$$

$$= \sum_{m,n} \left(X, \Phi^* ((E_\Lambda \xi_m) \otimes (F_\Xi \eta_n)^c) \right)_{\mathcal{C}_2(\mathcal{H})}$$

$$= \sum_{m,n} \left(X, \int_0^{\|H\|} \int_0^{\|K\|} \overline{\phi(s,t)} \, d(\pi_l(E_s)\pi_r(F_t))((E_\Lambda \xi_m) \otimes (F_\Xi \eta_n)^c) \right)_{\mathcal{C}_2(\mathcal{H})}$$

$$= \sum_{m,n} \left(X, \int_\Lambda \int_\Xi \overline{\phi(s,t)} \, (dE_s \xi_m) \otimes (dF_t \eta_n)^c \right)_{\mathcal{C}_2(\mathcal{H})}$$

$$= \sum_{m,n} \int_\Lambda \int_\Xi \phi(s,t) \, (X dE_s \xi_m, dF_t \eta_n)$$

$$= \int_\Lambda \int_\Xi \phi(s,t) k(s,t) \, d\lambda(s) \, d\mu(t)$$

because of

$$\sum_{m,n}(XE_\Lambda\xi_m, F_\Xi\eta_n) = (R\chi_\Lambda, \chi_\Xi)_{L^2(\mu)} = \int_\Lambda\int_\Xi k(s,t)\,d\lambda(s)\,d\mu(t).$$

We thus conclude that $\phi(s,t)k(s,t)$ is the kernel of the trace class operator $L^2(\lambda) \to L^2(\mu)$ corresponding to $\Phi(X)|_{\mathcal{H}_1} : \mathcal{H}_1 \to \mathcal{H}_2$.

Proof of (ii) $\Rightarrow$ (iii)

This is the most non-trivial part in Peller's theorem, and requires the notion of one-integrable operators (between Banach spaces) and the Grothendieck theorem. Assume that ϕ satisfies (ii) and define an integral operator $T_0 : L^1(\lambda) \to L^\infty(\mu)$ by

$$(T_0 f)(t) = \int_0^{\|H\|}\phi(s,t)f(s)\,d\lambda(s) \quad \text{for } f \in L^1(\lambda).$$

What we need to show is that T_0 falls into the operator ideal $\mathfrak{I}_1(L^1(\lambda), L^\infty(\mu))$ consisting of one-integral operators in the space of bounded operators $L^1(\lambda) \to L^\infty(\mu)$. Our standard reference for the theory on operator ideals on Banach spaces is Pietsch's textbook [72] (see especially [72, §19.2]).

It is known (see [72, 19.2.13]) that $\mathfrak{I}_1(L^1(\lambda), L^\infty(\mu))$ is dual to the space of compact operators $L^\infty(\mu) \to L^1(\lambda)$. Thanks to [72, 10.3.6 and E.3.1], to show $T_0 \in \mathfrak{I}_1(L^1(\lambda), L^\infty(\mu))$, it suffices to prove that there exists a constant C such that

$$|\mathrm{trace}(T_0 Q)| \le C\|Q\| \tag{2.9}$$

for finite-rank operators $Q : L^\infty(\mu) \to L^1(\lambda)$ of the form $Q = \sum_{k=1}^l \langle\,\cdot\,, h_k\rangle g_k$ with $g_k \in L^1(\lambda)$ and $h_k \in L^1(\mu)$. Here, $\langle\,\cdot\,,\cdot\,\rangle$ denotes the duality between $L^\infty(\mu)$ and $L^1(\mu)$ and

$$\mathrm{trace}(T_0 Q) = \sum_{k=1}^n\langle T_0 g_k, h_k\rangle$$

for $T_0 Q = \sum_{k=1}^l\langle\,\cdot\,, h_k\rangle T_0 g_k$.

To show (2.9), one may and do assume that g_k, h_k are finite linear combinations of characteristic functions, say $g_k = \sum_{i=1}^m \alpha_{ki}\chi_{\Lambda_i}$, $h_k = \sum_{j=1}^n \beta_{kj}\chi_{\Xi_j}$ where $\mathcal{A} = \{\Lambda_1, \ldots, \Lambda_m\}$ and $\mathcal{B} = \{\Xi_1, \ldots, \Xi_n\}$ are measurable partitions of $[0, \|H\|]$ and $[0, \|K\|]$ respectively. For $p = 1, 2, \infty$ write $L^p(\mathcal{A}, \lambda)$ for the (finite-dimensional) subspace of $L^p(\lambda)$ consisting of $\mathcal{A}$-measurable functions (i.e., linear combinations of χ_{Λ_i}'s) and $L^p(\mathcal{B}, \mu)$ similarly. The conditional expectation $E_\mathcal{B} : L^p(\mu) \to L^p(\mathcal{B}, \mu)$ is given by

$$E_\mathcal{B}f = \sum_{j=1}^n\mu(\Xi_j)^{-1}\left(\int_{\Xi_j}f\,d\mu\right)\chi_{\Xi_j}.$$

Set $\tilde{Q} = Q|_{L^\infty(\mathcal{B},\mu)} : L^\infty(\mathcal{B}, \mu) \to L^1(\mathcal{A}, \lambda)$ so that we have $Q = \tilde{Q} \circ E_\mathcal{B}$.

According to [61, Theorem 4.3] (based on the Grothendieck theorem) together with [61, Proposition 3.1], we see that $\tilde{Q}$ admits a factorization

$$L^\infty(\mathcal{B},\mu) \xrightarrow{\tilde{M}_2} L^2(\mathcal{B},\mu) \xrightarrow{\tilde{R}} L^1(\mathcal{A},\lambda), \tag{2.10}$$

where $\tilde{M}_2$ is the multiplication by a function $\tilde{\eta} \in L^2(\mathcal{B},\mu)$ and $\tilde{R}$ is an operator such that

$$\|\tilde{\eta}\|_{L^2(\mu)} = 1 \quad \text{and} \quad \|\tilde{R}\| \le K_G\|\tilde{Q}\| \tag{2.11}$$

with the Grothendieck constant K_G. Apply [61, Theorem 4.3] once again to the transpose $\tilde{R}^t : L^\infty(\mathcal{A},\lambda) \to L^2(\mathcal{B},\mu)$ to get the following factorization of $\tilde{R}^t$:

$$L^\infty(\mathcal{A},\lambda) \xrightarrow{\hat{M}_1} L^2(\mathcal{A},\lambda) \xrightarrow{\hat{S}} L^2(\mathcal{B},\mu),$$

where $\hat{M}_1$ is the multiplication by a function $\tilde{\xi} \in L^2(\mathcal{A},\lambda)$ and $\hat{S}$ is an operator such that

$$\|\tilde{\xi}\|_{L^2(\lambda)} = 1 \quad \text{and} \quad \|\hat{S}\| \le K_G\|\tilde{R}^t\| = K_G\|\tilde{R}\|. \tag{2.12}$$

Hence $\tilde{R}$ is factorized as

$$L^2(\mathcal{B},\mu) \xrightarrow{\tilde{S}=\hat{S}^t} L^2(\mathcal{A},\lambda) \xrightarrow{\tilde{M}_1=\hat{M}_1^t} L^1(\mathcal{A},\lambda), \tag{2.13}$$

where $\tilde{M}_1$ is again the multiplication by $\tilde{\xi}$. Combining (2.10) and (2.13) implies that Q is factorized as

$$L^\infty(\mu) \xrightarrow{E_\mathcal{B}} L^\infty(\mathcal{B},\mu) \xrightarrow{\tilde{M}_2} L^2(\mathcal{B},\mu) \xrightarrow{\tilde{S}} L^2(\mathcal{A},\lambda) \xrightarrow{\tilde{M}_1} L^1(\mathcal{A},\lambda) \hookrightarrow L^1(\lambda).$$

Let $S = \tilde{S}E_\mathcal{B} : L^2(\mu) \to L^2(\mathcal{B},\mu) \to L^2(\mathcal{A},\lambda) \subseteq L^2(\lambda)$ and $M_1 : L^2(\lambda) \to L^1(\lambda)$, $M_2 : L^\infty(\mu) \to L^2(\mu)$ be the multiplications by $\tilde{\xi}, \tilde{\eta}$ respectively. Since

$$Q = \tilde{M}_1\tilde{S}\tilde{M}_2E_\mathcal{B} = \tilde{M}_1\tilde{S}E_\mathcal{B}M_2 = M_1SM_2,$$

we finally obtain a factorization of Q as follows:

$$L^\infty(\mu) \xrightarrow{M_2} L^2(\mu) \xrightarrow{S} L^2(\lambda) \xrightarrow{M_1} L^1(\lambda)$$

with

$$\|S\| = \|\tilde{S}\| \le K_G\|\tilde{R}\| \le K_G^2\|\tilde{Q}\| = K_G^2\|Q\| \tag{2.14}$$

thanks to (2.11) and (2.12).

Notice that $M_2T_0M_1 : L^2(\lambda) \to L^2(\mu)$ is the integral operator

$$(M_2T_0M_1f)(t) = \int_0^{\|H\|} \phi(s,t)\tilde{\xi}(s)\tilde{\eta}(t)f(s)\,d\lambda(s).$$

Since $\tilde{\xi}(s)\tilde{\eta}(t)$ is obviously a kernel of a rank-one operator $L^2(\lambda) \to L^2(\mu)$, the assumption (ii) implies that $M_2T_0M_1$ is a trace class operator. Now, it is easy to see that

$$\text{trace}(T_0 Q) = \text{trace}(T_0 M_1 S M_2) = \text{Tr}(M_2 T_0 M_1 S) \tag{2.15}$$

with the (ordinary) trace Tr for the trace class operator $M_2 T_0 M_1 S$ on $L^2(\mu)$.

For every $\xi \in L^2(\lambda)$ and $\eta \in L^2(\mu)$, the assumption (ii) guarantees that one can define a trace class operator $A(\xi, \eta) : L^2(\lambda) \to L^2(\mu)$ by

$$(A(\xi,\eta)f)(t) = \int_0^{\|H\|} \phi(s,t)\xi(s)\eta(t)f(s)\,d\lambda(s);$$

in particular, $M_2 T_0 M_1 = A(\tilde{\xi}, \tilde{\eta})$. Write $\mathcal{C}_1(L^2(\lambda), L^2(\mu))$ for the Banach space (with trace norm $\|\cdot\|_{\mathcal{C}_1(L^2(\lambda),L^2(\mu))}$) consisting of trace class operators $L^2(\lambda) \to L^2(\mu)$.

Lemma 2.3. *There exists a constant $\tilde{C}$ such that*

$$\|A(\xi,\eta)\|_{\mathcal{C}_1(L^2(\lambda),L^2(\mu))} \leq \tilde{C}\|\xi\|_{L^2(\lambda)}\|\eta\|_{L^2(\mu)} \tag{2.16}$$

for each $\xi \in L^2(\lambda)$ and $\eta \in L^2(\mu)$.

Proof. For a fixed $\xi \in L^2(\lambda)$ let us consider the linear map

$$A(\xi,\cdot) : \eta \in L^2(\mu) \mapsto A(\xi,\eta) \in \mathcal{C}_1(L^2(\lambda), L^2(\mu)),$$

whose graph is shown to be closed. We assume

$$\eta_n \longrightarrow \eta \text{ in } L^2(\mu) \quad \text{and} \quad A(\xi,\eta_n) \longrightarrow B \text{ in } \mathcal{C}_1(L^2(\lambda), L^2(\mu)).$$

Choose and fix $f \in L^2(\lambda)$, and notice

$$\|A(\xi,\eta_n)f - Bf\|_{L^2(\mu)} \leq \|A(\xi,\eta_n) - B\|_{B(L^2(\lambda),L^2(\mu))}\|f\|_{L^2(\lambda)}$$
$$\leq \|A(\xi,\eta_n) - B\|_{\mathcal{C}_1(L^2(\lambda),L^2(\mu))}\|f\|_{L^2(\lambda)} \longrightarrow 0.$$

From these L^2-convergences, after passing to a subsequence if necessary, we may and do assume

$$\eta_n(t) \longrightarrow \eta(t) \quad \text{and} \quad (A(\xi,\eta_n)f)(t) \longrightarrow (Bf)(t) \quad \text{for } \mu\text{-a.e. } t.$$

We then estimate

$$|(A(\xi,\eta_n)f)(t) - (A(\xi,\eta)f)(t)|$$
$$\leq \left| \int_0^{\|H\|} \phi(s,t)(\eta_n(t) - \eta(t))\xi(s)f(s)d\lambda(s) \right|$$
$$\leq |\eta_n(t) - \eta(t)| \times \|\phi\|_\infty \times \int_0^{\|H\|} |\xi(s)f(s)|\,d\lambda(s).$$

The last integral here being finite (due to $\xi, f \in L^2(\lambda)$), we conclude $(Bf)(t) = (A(\xi,\eta)f)(t)$ for μ-a.e. t. This means $Bf = A(\xi,\eta)f \in L^2(\mu)$

and the arbitrariness of $f \in L^2(\lambda)$ shows $B = A(\xi, \eta)$ as desired. Therefore, the closed graph theorem guarantees the boundedness of $A(\xi, \cdot)$, i.e.,

$$\|A(\xi, \cdot)\| = \sup\{\|A(\xi, \eta)\|_{\mathcal{C}_1(L^2(\lambda), L^2(\mu))} : \eta \in L^2(\mu), \ \|\eta\|_{L^2(\mu)} \le 1\} < \infty,$$
$$\|A(\xi, \eta)\|_{\mathcal{C}_1(L^2(\lambda), L^2(\mu))} \le \|A(\xi, \cdot)\| \times \|\eta\|_{L^2(\mu)}. \tag{2.17}$$

We next consider the linear map

$$A : \ \xi \in L^2(\lambda) \mapsto A(\xi, \cdot) \in B(L^2(\mu), \mathcal{C}_1(L^2(\lambda), L^2(\mu))).$$

To show the closedness of the graph again, we assume

$$\xi_n \longrightarrow \xi \ \text{in} \ L^2(\lambda) \quad \text{and} \quad A(\xi_n, \cdot) \longrightarrow C \ \text{in} \ B(L^2(\mu), \mathcal{C}_1(L^2(\lambda), L^2(\mu))).$$

We need to show $A(\xi, \cdot) = C \in B(L^2(\mu), \mathcal{C}_1(L^2(\lambda), L^2(\mu)))$, i.e., $A(\xi, \eta) = C(\eta) \in \mathcal{C}_1(L^2(\lambda), L^2(\mu))$ $(\eta \in L^2(\mu))$. For each fixed $f \in L^2(\lambda)$ (and $\eta \in L^2(\mu)$), we have $A(\xi_n, \eta)f \to C(\eta)f$ in $L^2(\mu)$. From this L^2-convergence and the fact $\eta \in L^2(\mu)$, after passing to a subsequence, we have

$$(A(\xi_n, \eta)f)(t) \longrightarrow (C(\eta)f)(t) \quad \text{and} \quad |\eta(t)| < \infty \quad \text{for } \mu\text{-a.e. } t.$$

We estimate

$$|(A(\xi_n, \eta)f)(t) - A(\xi, \eta)f)(t)|$$
$$\le \left| \int_0^{\|H\|} \phi(s, t)\eta(t)(\xi_n(s) - \xi(s))f(s)d\lambda(s) \right|$$
$$\le \|\phi\|_\infty \times |\eta(t)| \times \int_0^{\|H\|} |(\xi_n(s) - \xi(s))f(s)| \ d\lambda(s)$$
$$\le \|\phi\|_\infty \times |\eta(t)| \times \|\xi_n - \xi\|_{L^2(\lambda)}\|f\|_{L^2(\lambda)}.$$

Therefore, we have $(A(\xi, \eta)f)(t) = (C(\eta)f)(t)$ for μ-a.e. t, showing $A(\xi, \eta)f = C(\eta)f \in L^2(\mu)$ $(f \in L^2(\lambda))$ and $A(\xi, \eta) = C(\eta) \in \mathcal{C}_1(L^2(\lambda), L^2(\mu))$ (for each $\eta \in L^2(\mu)$). Thus, the closed graph theorem shows the boundedness

$$\|A(\xi, \cdot)\| \le \tilde{C}\|\xi\|_{L^2(\lambda)} \quad \text{for some } \tilde{C},$$

which together with (2.17) implies the inequality (2.16). □

We are now ready to prove (iii). By combining the above estimates (2.15), (2.16), (2.11), (2.12) and (2.14) altogether, we get

$$|\text{trace}(T_0 Q)| \le \|A(\tilde{\xi}, \tilde{\eta})S\|_{\mathcal{C}_1(L^2(\mu))} \le \tilde{C}\|\tilde{\xi}\|_{L^2(\lambda)}\|\tilde{\eta}\|_{L^2(\mu)}\|S\| \le \tilde{C}K_G^2\|Q\|,$$

proving (2.9) with a constant $C = \tilde{C}K_G^2$ (independent of Q). Thus, $T_0 \in \mathfrak{I}_1(L^1(\lambda), L^\infty(\mu))$ is established.

The following fact is known among other characterizations (see [72, 19.2.6]): a bounded operator $T : L^1(\lambda) \to L^\infty(\mu)$ belongs to $\mathfrak{I}_1(L^1(\lambda), L^\infty(\mu))$ if and only if there exist a probability space (Ω, σ) and bounded operators $T_1 : L^1(\lambda) \to L^\infty(\Omega; \sigma)$, $T_2 : L^1(\Omega; \sigma) \to L^\infty(\mu)^{**}$ such that

$$
\begin{array}{ccc}
L^1(\lambda) & \xrightarrow{\ T\ } L^\infty(\mu) \hookrightarrow L^\infty(\mu)^{**} \\
T_1 \downarrow & \uparrow T_2 \\
L^\infty(\Omega; \sigma) & \hookrightarrow & L^1(\Omega; \sigma)
\end{array}
$$

is commutative. Therefore, we can factorize T_0 as follows:

$$
L^1(\lambda) \xrightarrow{\ T_1\ } L^\infty(\Omega; \sigma) \hookrightarrow L^1(\Omega; \sigma) \xrightarrow{\ T_2\ } L^\infty(\mu),
$$

where (Ω, σ) is a finite measure space and T_1, T_2 are bounded operators. Indeed, $L^\infty(\mu)$ is complemented in $L^\infty(\mu)^{**}$, and this T_2 is the composition of a projection map (actually a norm-one projection due to M. Hasumi's result in [35], and also see [76, p. 148, Exercise 22 and p. 299, Exercise 10]) $L^\infty(\mu)^{**} \to L^\infty(\mu)$ and the preceding $T_2 : L^1(\Omega; \sigma) \to L^\infty(\mu)^{**}$.

Thanks to Lemma 2.4 below applied to the preceding bounded operators T_1, T_2, there exist $\alpha \in L^\infty([0, \|H\|] \times \Omega; \lambda \times \sigma)$ and $\beta \in L^\infty([0, \|K\|] \times \Omega; \mu \times \sigma)$ such that

$$
(T_1 f)(x) = \int_0^{\|H\|} \alpha(s, x) f(s) \, d\lambda(s) \quad \text{for } f \in L^1(\lambda),
$$

$$
(T_2 g)(t) = \int_\Omega \beta(t, x) g(x) \, d\sigma(x) \quad \text{for } g \in L^1(\Omega, \sigma).
$$

Therefore, we have

$$
(T_0 f)(t) = \int_\Omega \int_0^{\|H\|} \alpha(s, x) \beta(t, x) f(s) \, d\lambda(s) \, d\sigma(x)
$$

$$
= \int_0^{\|H\|} \left(\int_\Omega \alpha(s, x) \beta(t, x) \, d\sigma(x) \right) f(s) \, d\lambda(s) \quad \text{for } f \in L^1(\lambda),
$$

which yields (iii) and the proof of Theorem 2.2 is completed.

The next result can be found in [47] as a corollary of a more general result (see [47, §XI.1, Theorem 6]), and a short direct proof is presented below for the reader's convenience.

Lemma 2.4. *Let (Ω_1, σ_1) and (Ω_2, σ_2) be finite measure spaces. For a given bounded operator $T : L^1(\Omega_1; \sigma_1) \to L^\infty(\Omega_2; \sigma_2)$ there exists a unique $\tau \in L^\infty(\Omega_1 \times \Omega_2; \sigma_1 \times \sigma_2)$ satisfying*

$$
(Tf)(y) = \int_{\Omega_1} \tau(x, y) f(x) \, d\sigma_1(x) \quad \text{for } f \in L^1(\Omega_1; \sigma_1).
$$

Proof. Choose and fix a measurable set $\Xi \subseteq \Omega_2$. For each $f \in L^1(\sigma_1)$ we observe the trivial estimate

$$|\langle Tf, \chi_\Xi \rangle_{\sigma_2}| \leq \sigma_2(\Xi) \times \|Tf\|_{L^\infty(\sigma_2)} \leq \sigma_2(\Xi) \times \|T\| \times \|f\|_{L^1(\sigma_1)}$$

(with the standard bilinear form $\langle \cdot, \cdot \rangle_{\sigma_2}$ giving rise to the duality between $L^\infty(\sigma_2)$ and $L^1(\sigma_2)$), showing the existence of $h_\Xi \in L^\infty(\Omega_1; \sigma_1)$ satisfying $\|h_\Xi\|_{L^\infty(\sigma_1)} \leq \|T\|$ and

$$\langle Tf, \chi_\Xi \rangle_{\sigma_2} = \sigma_2(\Xi) \times \langle h_\Xi, f \rangle_{\sigma_1} \quad \text{for } f \in L^1(\sigma_1).$$

Let Π denote the set of all finite measurable partitions of Ω_2, which is a directed set in the order of refinement. For every $\pi \in \Pi$ we set

$$\tau_\pi(x, y) = \sum_{\Xi \in \pi} h_\Xi(x) \chi_\Xi(y), \qquad (x, y) \in \Omega_1 \times \Omega_2,$$

so that a net $\{\tau_\pi\}_{\pi \in \Pi}$ in $L^\infty(\sigma_1 \times \sigma_2)$ satisfies $\|\tau_\pi\|_{L^\infty(\sigma_1 \times \sigma_2)} \leq \|T\|$ and

$$\langle Tf, \chi_\Xi \rangle_{\sigma_2} = \langle \tau_\pi, f \times \chi_\Xi \rangle_{\sigma_1 \times \sigma_2} \quad \text{for } f \in L^1(\sigma_1)$$

for each π-measurable Ξ (i.e., π refines $\{\Xi, \Omega_2 \setminus \Xi\}$). Thanks to the w*-compactness of $\{\phi \in L^\infty(\sigma_1 \times \sigma_2); \|\phi\|_{L^\infty(\sigma_1 \times \sigma_2)} \leq \|T\|\}$ one can take a w*-limit point τ of $\{\tau_\pi\}_{\pi \in \Pi}$. Then it is easy to see that

$$\langle Tf, \chi_\Xi \rangle_{\sigma_2} = \langle \tau, f \times \chi_\Xi \rangle_{\sigma_1 \times \sigma_2} = \int_\Xi \left(\int_{\Omega_1} \tau(x, y) f(x) \, d\sigma_1(x) \right) d\sigma_2(y)$$

for each $f \in L^1(\Omega_1; \sigma_1)$ and each measurable set $\Xi \subseteq \Omega_2$. This implies the desired integral expression, and the uniqueness of τ is obvious.   □

2.2 Extension to $B(\mathcal{H})$

We assume the condition (iv) in Theorem 2.2 (i.e., $\phi(s, t)$ admits the integral expression (2.3) with the finiteness condition described in (iv)) and will explain how to extend $\Phi(\cdot)$ to a bounded transformation on $B(\mathcal{H})$ by making use of the duality $B(\mathcal{H}) = \mathcal{C}_1(\mathcal{H})^*$ via

$$(X, Y) \in \mathcal{C}_1(\mathcal{H}) \times B(\mathcal{H}) \mapsto \text{Tr}(XY) \in \mathbf{C}.$$

To do so, we first note that the roles of the variables s, t (and those of dE_s and dF_t) are symmetric. Thus, the function

$$\tilde{\phi}(t, s) = \phi(s, t) = \int_\Omega \beta(t, x) \alpha(s, x) \, d\sigma(x)$$

gives rise to the following transformation on $\mathcal{C}_1(\mathcal{H})$:

$$\tilde{\Phi}(X) = \int_0^{\|K\|} \int_0^{\|H\|} \tilde{\phi}(t, s) \, dF_t X \, dE_s.$$

We consider its transpose $\tilde{\Phi}^t$ on $B(\mathcal{H}) = \mathcal{C}_1(\mathcal{H})^*$, that is,

$$\mathrm{Tr}(X\tilde{\Phi}^t(Y)) = \mathrm{Tr}(\tilde{\Phi}(X)Y) \quad \text{for } X \in \mathcal{C}_1(\mathcal{H}), \ Y \in B(\mathcal{H}). \tag{2.18}$$

Let us take $X = \xi \otimes \eta^c$ here. Then, the left side of (2.18) is obviously the inner product $(\tilde{\Phi}^t(Y)\xi, \eta)$. On the other hand, we have

$$\tilde{\Phi}(X) = \int_\Omega \tilde{\xi}(x) \otimes \tilde{\eta}(x)^c \, d\sigma(x)$$

with

$$\tilde{\xi}(x) = \int_0^{\|K\|} \beta(t, x) \, dF_t \xi \quad \text{and} \quad \tilde{\eta}(x) = \int_0^{\|H\|} \overline{\alpha(s, x)} \, dE_s \eta \tag{2.19}$$

(see (2.4), but recall that the roles of α and β were switched). We claim that the right side of (2.18) (when $X = \xi \otimes \eta^c$) is $\int_\Omega (Y\tilde{\xi}(x), \tilde{\eta}(x))d\sigma(x)$. In fact, for vectors ξ', η' we have

$$(\tilde{\Phi}(X)Y\xi', \eta') = \int_\Omega (Y\xi', \tilde{\eta}(x))(\tilde{\xi}(x), \eta') \, d\sigma(x)$$

$$= \int_\Omega (\tilde{\xi}(x), \eta')(\xi', Y^*\tilde{\eta}(x)) \, d\sigma(x)$$

thanks to (2.5). Let $\{e_n\}_{n=1,2,\cdots}$ be an orthonormal basis for $\mathcal{H}$. Since $\tilde{\Phi}(X)Y \in \mathcal{C}_1(\mathcal{H})$, from the preceding expression we get

$$\mathrm{Tr}(\tilde{\Phi}(X)Y) = \sum_{n=1}^\infty (\tilde{\Phi}(X)Ye_n, e_n) = \sum_{n=1}^\infty \int_\Omega (\tilde{\xi}(x), e_n)(e_n, Y^*\tilde{\eta}(x)) \, d\sigma(x)$$

(see [29, Chapter III, §8]). Here, we would like to switch the order of $\sum_{n=1}^\infty$ and $\int_\Omega$, which is guaranteed by the Fubini theorem thanks to the following integrability estimate:

$$\int_\Omega \sum_{n=1}^\infty |(\tilde{\xi}(x), e_n)(e_n, Y^*\tilde{\eta}(x))| \, d\sigma(x)$$

$$\leq \int_\Omega \left(\sum_{n=1}^\infty |(\tilde{\xi}(x), e_n)|^2 \right)^{1/2} \left(\sum_{n=1}^\infty |(e_n, Y^*\tilde{\eta}(x))|^2 \right)^{1/2} d\sigma(x)$$

$$= \int_\Omega \|\tilde{\xi}(x)\| \times \|Y^*\tilde{\eta}(x)\| \, d\sigma(x) \leq \|Y\| \int_\Omega \|\tilde{\xi}(x)\| \times \|\tilde{\eta}(x)\| \, d\sigma(x) < \infty$$

(see (2.6) and (2.7)). Hence, we get

$$\mathrm{Tr}(\tilde{\Phi}(X)Y) = \int_\Omega \sum_{n=1}^\infty (\tilde{\xi}(x), e_n)(e_n, Y^*\tilde{\eta}(x)) \, d\sigma(x)$$

$$= \int_\Omega (\tilde{\xi}(x), Y^*\tilde{\eta}(x)) \, d\sigma(x) = \int_\Omega (Y\tilde{\xi}(x), \tilde{\eta}(x)) \, d\sigma(x).$$

Therefore, the claim has been proved, and (for $X = \xi \otimes \eta^c$) (2.18) means

$$(\tilde{\Phi}^t(Y)\xi, \eta) = \int_\Omega (Y\tilde{\xi}(x), \tilde{\eta}(x))\, d\sigma(x) \tag{2.20}$$

with the vectors $\tilde{\xi}(x)$ and $\tilde{\eta}(x)$ defined by (2.19).

When $Y = \xi' \otimes \eta'^c$, the right side of (2.20) is

$$\int_\Omega (\tilde{\xi}(x), \eta')(\xi', \tilde{\eta}(x))\, d\sigma(x)$$

$$= \int_\Omega \left(\int_0^{\|K\|} \beta(t,x)dF_t\xi, \eta' \right) \left(\xi', \int_0^{\|H\|} \overline{\alpha(s,x)}dE_s\eta \right) d\sigma(x)$$

$$= \int_\Omega \left(\xi, \int_0^{\|K\|} \overline{\beta(t,x)}dF_t\eta' \right) \left(\int_0^{\|H\|} \alpha(s,x)dE_s\xi', \eta \right) d\sigma(x)$$

$$= \int_\Omega (Y(x)\xi, \eta)\, d\sigma(x)$$

with the rank-one operator

$$Y(x) = \left(\int_0^{\|H\|} \alpha(s,x)\, dE_s\xi' \right) \otimes \left(\int_0^{\|K\|} \overline{\beta(t,x)}\, dF_t\eta' \right)^c.$$

But, notice that the two involved vectors here are exactly those defined from ξ' and η' according to the formula (2.4). Therefore, we have shown

$$\tilde{\Phi}^t(Y) = \int_\Omega Y(x)\, d\sigma(x) = \Phi(Y) \tag{2.21}$$

for a rank-one (and hence finite-rank) operator Y.

For a general Hilbert-Schmidt class operator Y, we choose a sequence $\{Y_n\}_{n=1,2,\dots}$ of finite-rank operators tending to Y in $\|\cdot\|_2$. Since the convergence is also valid in the operator norm and $\tilde{\Phi}^t$ (being defined as a transpose) is bounded relative to the operator norm, we have $\tilde{\Phi}^t(Y) = \|\cdot\|\text{-}\lim_{n\to\infty} \tilde{\Phi}^t(Y_n)$. On the other hand, we know

$$\Phi(Y) = \|\cdot\|_2\text{-}\lim_{n\to\infty} \Phi(Y_n) = \|\cdot\|_2\text{-}\lim_{n\to\infty} \tilde{\Phi}^t(Y_n)$$

thanks to (2.1) and (2.21). Therefore, we conclude $\tilde{\Phi}^t(Y) = \Phi(Y)$ so that $\tilde{\Phi}^t$ is indeed an extension of Φ (originally defined on $\mathcal{C}_2(\mathcal{H})$).

The discussions so far justify the use of the notation $\Phi(Y)$ (for $Y \in B(\mathcal{H})$) for expressing $\tilde{\Phi}^t(Y)$, and we shall also use the symbolic notation

$$\Phi(Y)\ (= \Phi_{H,K}(Y)) = \int_0^{\|H\|} \int_0^{\|K\|} \phi(s,t)\, dE_s Y dF_t \quad \text{(for } Y \in B(\mathcal{H}))$$

in the rest of the monograph.

Remark 2.5.

(i) The map $\Phi : X \in B(\mathcal{H}) \mapsto \Phi(X) \in B(\mathcal{H})$ is always w*-w*-continuous (i.e., $\sigma(B(\mathcal{H}), \mathcal{C}_1(\mathcal{H}))$-$\sigma(B(\mathcal{H}), \mathcal{C}_1(\mathcal{H}))$-continuous) because it was defined as the transpose of the bounded transformation $\tilde{\Phi}$ on $\mathcal{C}_1(\mathcal{H})$.

(ii) From (2.19) and (2.20) we observe

$$(\Phi(Y)\xi, \eta) = \int_\Omega (Y\beta(K, x)\xi, \alpha(H, x)^*\eta) \, d\sigma(x)$$

$$= \int_\Omega (\alpha(H, x)Y\beta(K, x)\xi, \eta) \, d\sigma(x)$$

with the usual function calculus

$$\alpha(H, x) = \int_0^{\|H\|} \alpha(H, x) \, dE_s \quad \text{and} \quad \beta(K, x) = \int_0^{\|K\|} \beta(t, x) \, dF_t.$$

Therefore, $\Phi(X)$ (for $X \in B(\mathcal{H})$) can be simply written as the integral

$$\Phi(X) = \int_\Omega \alpha(H, x)X\beta(K, x) \, d\sigma(x)$$

in the weak sense. Remark that the integral expression (2.3) for $\varphi(s, t)$ is far from being unique. Nevertheless, there is no ambiguity for the definition of $\Phi(X)$. Indeed, the definition of $\tilde{\Phi}(X)$ $(= \tilde{\Phi}|_{\mathcal{C}_1(\mathcal{H})}(X))$ for $X \in \mathcal{C}_1(\mathcal{H})$ $(\subseteq \mathcal{C}_2(\mathcal{H}))$ does not depend on this expression (see (2.2)), and $\Phi(X)$ (for $X \in B(\mathcal{H}) = \mathcal{C}(\mathcal{H})^*$) was defined as the transpose.

(iii) From the expression in (ii) we obviously have

$$f(H)(\Phi(X))g(K) = \Phi(f(H)Xg(K))$$

for all bounded Borel functions f, g.

2.3 Norm estimates

We begin by investigating a relationship between the two norms

$$\|\Phi\|_{(\infty,\infty)} = \sup\{\|\Phi(X)\| : \|X\| \leq 1\},$$
$$\|\Phi\|_{(1,1)} = \sup\{\|\Phi(X)\|_1 : \|X\|_1 \leq 1\}.$$

To do so, besides Φ and $\tilde{\Phi}$ we also make use of the following auxiliary double integral operator:

$$\bar{\Phi}(X) = \int_0^{\|H\|} \int_0^{\|K\|} \overline{\phi(s, t)} \, dE_s X \, dF_t.$$

Proposition 2.6. (M. Sh. Birman and M. Z. Solomyak, [16]) *For a Schur multiplier $\phi \in L^\infty([0, \|H\|] \times [0, \|K\|]; \lambda \times \mu)$ we have*

$$\|\Phi\|_{(1,1)} = \|\Phi\|_{(\infty,\infty)}.$$

Proof. For $X \in \mathcal{C}_2(\mathcal{H})$ we easily observe $\bar{\Phi}(X^*)^* = \tilde{\Phi}(X)$ and hence $\|\tilde{\Phi}\|_{(1,1)} = \|\bar{\Phi}\|_{(1,1)}$ by restricting the both sides to $\mathcal{C}_1(\mathcal{H})$ ($\subseteq \mathcal{C}_2(\mathcal{H})$). On the other hand, $\|\Phi\|_{(\infty,\infty)} = \|\tilde{\Phi}\|_{(1,1)}$ is obvious from the definition, i.e., Φ was defined as a transpose. Therefore, to prove the proposition it suffices to see $\|\Phi\|_{(1,1)} = \|\bar{\Phi}\|_{(1,1)}$.

One expresses $\mathcal{H}$ and E in the direct integral form as follows:

$$\mathcal{H} = \int_{[0,\|H\|]}^{\oplus} \mathcal{H}(s)\, d\lambda(s), \quad E_\Lambda = \int_{[0,\|H\|]}^{\oplus} \chi_\Lambda(s) 1_{\mathcal{H}(s)}\, d\lambda(s)$$

for Borel sets $\Lambda \subseteq [0, \|H\|]$. Note that it is the central decomposition of the von Neumann algebra $\{E_\Lambda : \Lambda \subseteq [0, \|H\|]\}'$ over its center

$$\{E_\Lambda : \Lambda \subseteq [0, \|H\|]\}'' \cong L^\infty([0, \|H\|]; \lambda).$$

(See [17, Chapter 7, §2] for more "operator-theoretical description".) Similarly, one can write

$$\mathcal{H} = \int_{[0,\|K\|]}^{\oplus} \tilde{\mathcal{H}}(t)\, d\mu(t), \quad F_\Xi = \int_{[0,\|K\|]}^{\oplus} \chi_\Xi(t) 1_{\tilde{\mathcal{H}}(t)}\, d\mu(t)$$

for Borel sets $\Xi \subseteq [0, \|K\|]$. A standard argument in the theory of direct integral shows that $\mathcal{C}_2(\mathcal{H})$ is represented as the direct integral

$$\mathcal{C}_2(\mathcal{H}) = \int_{[0,\|H\|]\times[0,\|K\|]}^{\oplus} \mathcal{C}_2(\mathcal{H}(s), \tilde{\mathcal{H}}(t))\, d(\lambda \times \mu)(s,t)$$

with the Hilbert-Schmidt class operators $\mathcal{C}_2(\mathcal{H}(s), \tilde{\mathcal{H}}(t))$ from $\mathcal{H}(s)$ into $\tilde{\mathcal{H}}(t)$. Take an $X = \int_{[0,\|H\|]\times[0,\|K\|]}^{\oplus} X(s,t)\, d(\lambda \times \mu)(s,t)$ in $\mathcal{C}_2(\mathcal{H})$. Since

$$E_\Lambda X F_\Xi = \int_{[0,\|H\|]\times[0,\|K\|]}^{\oplus} \chi_{\Lambda \times \Xi}(s,t) X(s,t)\, d(\lambda \times \mu)(s,t)$$

for Borel sets $\Lambda \subseteq [0, \|H\|]$ and $\Xi \subseteq [0, \|K\|]$, it is immediate to see that $\Phi(X)$ and $\bar{\Phi}(X)$ are written as

$$\Phi(X) = \int_{[0,\|H\|]\times[0,\|K\|]}^{\oplus} \phi(s,t) X(s,t)\, d(\lambda \times \mu)(s,t),$$

$$\bar{\Phi}(X) = \int_{[0,\|H\|]\times[0,\|K\|]}^{\oplus} \overline{\phi(s,t)} X(s,t)\, d(\lambda \times \mu)(s,t)$$

respectively. The measurable cross-section theorem guarantees that one can select measurable fields

$$\{J(s) : s \in [0, \|H\|]\} \quad \text{and} \quad \{\tilde{J}(s) : s \in [0, \|K\|]\}$$

of (conjugate linear) involutions $J(s) : \mathcal{H}(s) \to \mathcal{H}(s)$, $\tilde{J}(s) : \tilde{\mathcal{H}}(s) \to \tilde{\mathcal{H}}(s)$, and they give rise to the global involutions

$$J = \int_{[0,\|H\|]}^{\oplus} J(s)\, d\lambda(s) \quad \text{and} \quad \tilde{J} = \int_{[0,\|K\|]}^{\oplus} \tilde{J}(t)\, d\mu(t)$$

on the Hilbert space $\mathcal{H}$. Then we observe

$$\tilde{J}\Phi(X)J = \int_{[0,\|H\|]\times[0,\|K\|]}^{\oplus} \overline{\phi(s,t)}\tilde{J}(t)X(s,t)J(s)\, d(\lambda\times\mu)(s,t) = \bar{\Phi}(\tilde{J}XJ).$$

Since the map $X \mapsto \tilde{J}XJ$ is obviously isometric on $\mathcal{C}_1(\mathcal{H})$, the equality $\|\Phi\|_{(1,1)} = \|\bar{\Phi}\|_{(1,1)}$ is now obvious and the proposition has been proved. □

For each unitarily invariant norm $||| \cdot |||$, let $\mathcal{I}_{|||\cdot|||}$ and $\mathcal{I}_{|||\cdot|||}^{(0)}$ be the associated symmetrically normed ideals, that is,

$$\mathcal{I}_{|||\cdot|||} = \{X \in B(\mathcal{H}) : |||X||| < \infty\},$$
$$\mathcal{I}_{|||\cdot|||}^{(0)} = \text{the } ||| \cdot |||\text{-closure of } \mathcal{I}_{\mathrm{fin}} \text{ in } \mathcal{I}_{|||\cdot|||},$$

where $\mathcal{I}_{\mathrm{fin}}$ is the ideal of finite-rank operators (see [29, 37, 77] for details). For a Schur multiplier $\phi(t,s)$ we have shown

$$\|\Phi(X)\|_1 \leq k\|X\|_1 \ (X \in \mathcal{C}_1(\mathcal{H})) \text{ and } \|\Phi(X)\| \leq k\|X\| \ (X \in B(\mathcal{H})) \quad (2.22)$$

with $k = \|\Phi\|_{(1,1)} = \|\Phi\|_{(\infty,\infty)}$ (Proposition 2.6). The next result says that $\phi(s,t)$ is automatically a "Schur multiplier for all operator ideals $\mathcal{I}_{|||\cdot|||}$, $\mathcal{I}_{|||\cdot|||}^{(0)}$" with the same bound for

$$\|\Phi\|_{(|||\cdot|||,\,|||\cdot|||)} = \sup\{|||\Phi(X)||| : |||X||| \leq 1\}.$$

Proposition 2.7. Let $\phi(s,t)$ be a Schur multiplier with

$$\kappa = \|\Phi\|_{(1,1)} = \|\Phi\|_{(\infty,\infty)} \ (< \infty).$$

For any unitarily invariant norm $||| \cdot |||$ we have

$$|||\Phi(X)||| \leq \kappa|||X||| \ (\leq \infty)$$

for all $X \in B(\mathcal{H})$ so that Φ maps $\mathcal{I}_{|||\cdot|||}$ into itself. Moreover, Φ also maps the separable operator ideal $\mathcal{I}_{|||\cdot|||}^{(0)}$ into itself. In particular, $\Phi(X)$ is a compact operator as long as X is.

Proof. Recall the following expression for the Ky Fan norm as a K-functional:

$$|||X|||_{(n)} = \sum_{k=1}^{n} \mu_k(X)$$
$$= \inf\{n\|X_0\| + \|X_1\|_1 : X = X_0 + X_1\} \qquad (n = 1, 2, \cdots),$$

where $\{\mu_k(\cdot)\}_{k=1,2,\cdots}$ denotes the singular numbers (see [26, p. 289] for example). This expression together with (2.22) clearly shows $|||\Phi(X)|||_{(n)} \leq \kappa|||X|||_{(n)}$ for each n, which is known to be equivalent to the validity of $|||\Phi(X)||| \leq \kappa|||X|||$ for each unitarily invariant norm (see [37, Proposition 2.10]).

It remains to show $\Phi\left(\mathcal{I}_{|||\cdot|||}^{(0)}\right) \subseteq \mathcal{I}_{|||\cdot|||}^{(0)}$. When X is a finite-rank operator, $\Phi(X)$ is of trace class and can be approximated by a sequence $\{Y_n\}_{n=1,2,\cdots}$ of finite-rank operators in the $\|\cdot\|_1$-norm. Notice

$$|||\Phi(X) - Y_n||| \leq \|\Phi(X) - Y_n\|_1 \longrightarrow 0,$$

showing $\Phi(X) \in \mathcal{I}_{|||\cdot|||}^{(0)}$. For a general $X \in \mathcal{I}_{|||\cdot|||}^{(0)}$, one chooses a sequence $\{X_n\}_{n=1,2,\cdots}$ of finite-rank operators satisfying $\lim_{n\to\infty} |||X - X_n||| = 0$. Since $\Phi(X_n) \in \mathcal{I}_{|||\cdot|||}^{(0)}$ is already shown, the estimate $|||\Phi(X) - \Phi(X_n)||| \leq \kappa|||X - X_n||| \to 0$ (as $n \to \infty$) guarantees $\Phi(X) \in \mathcal{I}_{|||\cdot|||}^{(0)}$. $\square$

2.4 Technical results

Here we collect technical results. When we deal with integral expressions of means of operators in later chapters, a careful handling for supports of relevant operators will be required and some lemmas are prepared for this purpose. In the sequel we will denote the support projection of H by s_H.

Lemma 2.8. *Let ϕ, ψ be Schur multipliers (relative to (H, K)) with the corresponding double integral transformations Φ, Ψ respectively. Then, the pointwise product $\phi(s,t)\psi(s,t)$ is also a Schur multiplier, and the corresponding double integral transformation is the composition $\Phi \circ \Psi \ (= \Psi \circ \Phi)$.*

Proof. As in Theorem 2.2, (iv) we can write

$$\phi(s,t) = \int_\Omega \alpha(s,x)\beta(t,x) \, d\sigma(x),$$
$$\psi(s,t) = \int_{\Omega'} \alpha'(s,y)\beta'(t,y) \, d\sigma'(y).$$

We consider the product space $\Omega \times \Omega'$ equipped with the product measure $\sigma \times \sigma'$, and set

$$a : (s, x, y) \in [0, \|H\|] \times \Omega \times \Omega' \mapsto \alpha(s, x)\alpha'(s, y),$$
$$b : (t, x, y) \in [0, \|K\|] \times \Omega \times \Omega' \mapsto \beta(t, x)\beta'(t, y).$$

At first we note

$$\int_{\Omega \times \Omega'} |a(s, x, y)|^2 d(\sigma \times \sigma')(x, y) = \int_{\Omega} |\alpha(s, x)|^2 d\sigma(x) \times \int_{\Omega'} |\alpha'(s, y)|^2 d\sigma'(y)$$
$$\leq \left\| \int_{\Omega} |\alpha(\cdot, x)|^2 d\sigma(x) \right\|_{L^\infty(\lambda)} \times \left\| \int_{\Omega'} |\alpha'(\cdot, y)|^2 d\sigma'(y) \right\|_{L^\infty(\lambda)}$$

(and the similar estimate for b). Secondly, the Cauchy-Schwarz inequality implies

$$\int_{\Omega \times \Omega'} |a(s, x, y)b(t, x, y)| \, d(\sigma \times \sigma')(x, y)$$
$$\leq \left(\int_{\Omega \times \Omega'} |a(s, x, y)|^2 d(\sigma \times \sigma')(x, y) \right)^{1/2}$$
$$\times \left(\int_{\Omega \times \Omega'} |b(t, x, y)|^2 d(\sigma \times \sigma')(x, y) \right)^{1/2}.$$

From the two estimates we see the $\sigma \times \sigma'$-integrability of $a(s, x, y)b(t, x, y)$, and the Fubini theorem clearly shows

$$\int_{\Omega \times \Omega'} a(s, x, y)b(t, x, y) \, d(\sigma \times \sigma')(x, y)$$
$$= \int_{\Omega} \alpha(s, x)\beta(t, x) \, d\sigma(x) \times \int_{\Omega'} \alpha'(s, y)\beta'(t, y) \, d\sigma'(y) = \phi(s, t)\psi(s, t).$$

Therefore, the conditions stated in Theorem 2.2, (iv) have been checked for the product $\phi(s, t)\psi(s, t)$, and it is indeed a Schur multiplier.

Let Π be the double integral transformation corresponding to $\phi(s, t)\psi(s, t)$. Then, it is straight-forward to see $\Pi(X) = \Phi(\Psi(X))$ for each rank-one (hence finite-rank) operator X. Let $\{p_n\}_{n=1,2,\cdots}$ be a sequence of finite-rank projections tending to 1 in the strong operator topology. Then, for each $X \in B(\mathcal{H})$ the sequence $\{p_n X p_n\}$ tends to X strongly and hence in the $\sigma(B(\mathcal{H}), \mathcal{C}_1(\mathcal{H}))$-topology (because of $\|p_n X p_n\| \leq \|X\|$). Since $\Pi(p_n X p_n) = \Phi(\Psi(p_n X p_n))$ as remarked above, by letting $n \to \infty$ here, we conclude $\Pi(X) = \Phi(\Psi(X))$ due to the continuity stated in Remark 2.5, (i). $\square$

The additive version (which is much easier) is also valid. Namely, when ϕ, ψ are Schur multipliers, then so is the sum $\phi(s, t) + \psi(s, t)$ and the corresponding double integral transformation sends X to $\Phi(X) + \Psi(X)$.

Lemma 2.9. Let $\phi(s, t)$ be a Schur multiplier (relative to (H, K)) with the corresponding double integral transformation Φ. With the support projections s_H, s_K of H, K we have $s_H(\Phi(X))s_K = \Phi(s_H X s_K)$ and

$$\Phi(X) = s_H \Phi(X) s_K + \phi(H, 0) s_H X (1 - s_K) + (1 - s_H) X s_K \phi(0, K)$$
$$+ \phi(0, 0)(1 - s_H) X (1 - s_K).$$

Proof. The equation $s_H(\Phi(X)) s_K = \Phi(s_H X s_K)$ is seen from Remark 2.5, (iii). Recall the following expression mentioned in Remark 2.5, (ii):

$$\Phi(X) = \int_\Omega \alpha(H, x) X \beta(K, x) \, d\sigma(x)$$

in the weak sense. Since

$$\alpha(H, x) = \alpha(H, x) s_H + \alpha(0, x)(1 - s_H),$$
$$\beta(K, x) = \beta(K, x) s_K + \beta(0, x)(1 - s_K),$$

we have

$$\alpha(H, x) X \beta(K, x)$$
$$= \alpha(H, x) s_H X s_K \beta(K, x)$$
$$+ \beta(0, x) \alpha(H, x) s_H X (1 - s_K) + \alpha(0, x)(1 - s_H) X s_K \beta(K, x)$$
$$+ \alpha(0, x) \beta(0, x)(1 - s_H) X (1 - s_K).$$

The integration of the first term over Ω is $\Phi(s_H X s_K)$. The second term gives us

$$\int_\Omega \beta(0, x) \alpha(H, x) s_H X (1 - s_K) \, d\sigma(x)$$
$$= \left(\int_\Omega \int_0^{\|H\|} \alpha(s, x) \beta(0, x) \, dE_s \, d\sigma(x) \right) s_H X (1 - s_K)$$
$$= \left(\int_0^{\|H\|} \int_\Omega \alpha(s, x) \beta(0, x) \, d\sigma(x) \, dE_s \right) s_H X (1 - s_K)$$
$$= \left(\int_0^{\|H\|} \phi(s, 0) \, dE_s \right) s_H X (1 - s_K) = \phi(H, 0) s_H X (1 - s_K).$$

Of course the third term admits a similar integration. The last term gives us

$$\int_\Omega \alpha(0, x) \beta(0, x)(1 - s_H) X (1 - s_K) \, d\sigma(x)$$
$$= \left(\int_\Omega \alpha(0, x) \beta(0, x) \, d\sigma(x) \right) (1 - s_H) X (1 - s_K)$$
$$= \phi(0, 0)(1 - s_H) X (1 - s_K).$$

The above estimates altogether yield the desired expression for $\Phi(X)$.   □

We can consider $s_H(\Phi_{H,K}(X))s_K$ as an operator from $s_K\mathcal{H}$ to $s_H\mathcal{H}$, and denote it by $\Phi_{Hs_H,Ks_K}(s_HXs_K)$. It is possible to justify this (symbolic) notation by making use of double integral transformation for operators between two different spaces. The above lemma actually shows

$$s_H(\Phi_{H,K}(X))s_K = \Phi_{Hs_H,Ks_K}(s_HXs_K),$$
$$s_H(\Phi_{H,K}(X))(1-s_K) = s_H\phi(H,0)X(1-s_K),$$
$$(1-s_H)(\Phi_{H,K}(X))s_K = (1-s_H)X\phi(0,K)s_K,$$
$$(1-s_H)(\Phi_{H,K}(X))(1-s_K) = \phi(0,0)(1-s_H)X(1-s_K).$$

When dealing with means in later chapters we will mainly use Schur multipliers satisfying $\phi(s,0) = \phi(0,s) = bs$ $(s \geq 0)$ for some constant $b \geq 0$. Then, the expression in Lemma 2.9 becomes

$$\Phi_{H,K}(X) = s_H(\Phi_{H,K}(X))s_K + b\left(HX(1-s_K) + (1-s_H)XK\right) \qquad (2.23)$$

thanks to

$$\phi(H,0)s_H = bHs_H = bH, \quad \phi(0,K)s_K = bKs_K = bK \text{ and } \phi(0,0) = 0.$$

We fix signed measures ν_k $(k = 1,2,3)$ on the real line $\mathbf{R}$ with finite total variation and also a scalar a. With the Fourier transforms of these measures we set a bounded function π on $[0,\infty) \times [0,\infty)$ as

$$\pi(s,t) = \begin{cases} \hat{\nu}_1(\log s - \log t) & \text{if } s,t > 0, \\ \hat{\nu}_2(\log s) & \text{if } s > 0 \text{ and } t = 0, \\ \hat{\nu}_3(-\log t) & \text{if } s = 0 \text{ and } t > 0, \\ a & \text{if } s = t = 0. \end{cases}$$

Lemma 2.10. *The above $\pi(s,t)$ is a Schur multiplier for any pair (H,K) of positive operators, and the corresponding double integral transformation Π is given by*

$$\Pi(X) = \int_{-\infty}^{\infty} (Hs_H)^{ix}X(Ks_K)^{-ix}d\nu_1(x)$$
$$+ \int_{-\infty}^{\infty} (Hs_H)^{ix}X(1-s_K)\,d\nu_2(x)$$
$$+ \int_{-\infty}^{\infty} (1-s_H)X(Ks_K)^{-ix}d\nu_3(x)$$
$$+a(1-s_H)X(1-s_K).$$

We give a few remarks before proving the lemma. In the above expression, $(Hs_H)^{ix}$ for instance denotes a unitary operator on $s_H\mathcal{H}$ and it is zero on the orthogonal complement $(1-s_H)\mathcal{H}$, i.e., $(Hs_H)^{ix} = (Hs_H)^{ix}s_H$. We will mainly use this lemma (as well as the next Proposition 2.11) in the following special circumstances:

$\pi(s,0) = \pi(0,t) = c$ $(s > 0,\ t > 0)$ for some constant c and $\pi(0,0) = 0$.

This means $\nu_2 = \nu_3 = c\delta_0$ and $a = 0$, and hence in this case the expression in the lemma simply becomes

$$\Pi(X) = \int_{-\infty}^{\infty} (Hs_H)^{ix} X (Ks_K)^{-ix} d\nu_1(x)$$
$$+ c(s_H X(1 - s_K) + (1 - s_H)Xs_K).$$

Proof. We decompose the domain $\{(s,t) : s,t \geq 0\}$ into the four regions

$$\{(s,t) : s,t > 0\}, \quad \{(s,t) : s > 0,\ t = 0\},$$
$$\{(s,t) : s = 0, t > 0\}, \quad \{(s,t) : s = t = 0\}.$$

We accordingly set

$$\pi_1(s,t) = \begin{cases} \pi(s,t) & \text{if } s,t > 0, \\ 0 & \text{otherwise,} \end{cases} \qquad \pi_2(s,t) = \begin{cases} \pi(s,0) & \text{if } s > 0 \text{ and } t = 0, \\ 0 & \text{otherwise,} \end{cases}$$
$$\pi_3(s,t) = \begin{cases} \pi(0,t) & \text{if } s = 0 \text{ and } t > 0, \\ 0 & \text{otherwise,} \end{cases} \qquad \pi_4(s,t) = \begin{cases} \pi(0,0) & \text{if } s = t = 0, \\ 0 & \text{otherwise.} \end{cases}$$

So $\pi(s,t) = \sum_{k=1}^{4} \pi_k(s,t)$ is valid. We consider the following functions on $\mathbf{R}_+ \times \mathbf{R}$:

$$\alpha_1(s,x) = \begin{cases} s^{ix} \frac{d\nu_1}{d|\nu_1|}(x) & \text{if } s > 0, \\ 0 & \text{if } s = 0, \end{cases} \qquad \beta_1(t,x) = \begin{cases} t^{-ix} & \text{if } t > 0, \\ 0 & \text{if } t = 0, \end{cases}$$

$$\alpha_2(s,x) = \begin{cases} s^{ix} \frac{d\nu_2}{d|\nu_2|}(x) & \text{if } s > 0, \\ 0 & \text{if } s = 0, \end{cases} \qquad \beta_2(t,x) = \begin{cases} 0 & \text{if } t > 0, \\ 1 & \text{if } t = 0, \end{cases}$$

$$\alpha_3(s,x) = \begin{cases} 0 & \text{if } s > 0, \\ 1 & \text{if } s = 0, \end{cases} \qquad \beta_3(t,x) = \begin{cases} t^{-ix} \frac{d\nu_3}{d|\nu_3|}(x) & \text{if } t > 0, \\ 0 & \text{if } t = 0, \end{cases}$$

$$\alpha_4(s) = \begin{cases} 0 & \text{if } s > 0, \\ a & \text{if } s = 0, \end{cases} \qquad \beta_4(t) = \begin{cases} 0 & \text{if } t > 0, \\ 1 & \text{if } t = 0. \end{cases}$$

Here, $\frac{d\nu_k}{d|\nu_k|}(x)$ denotes the Radon-Nikodym derivative relative to the absolute value $|\nu_k|$. It is plain to observe

$$\pi_k(s,t) = \int_{-\infty}^{\infty} \alpha_k(s,x)\beta_k(t,x)\, d|\nu_k|(x) \quad \text{(for } k = 1,2,3)$$

and also $\pi_4(s,t) = \alpha_4(s)\beta_4(t)$. The finiteness condition in Theorem 2.2, (iv) is obviously satisfied (since $\frac{d\nu_k}{d|\nu_k|}$'s are bounded functions and $|\nu_k|$'s are finite measures) so that all π_k's are Schur multipliers. Thus, so is the sum π as was mentioned in the paragraph right after Lemma 2.8.

We begin with π_1 (with the corresponding double integral transformation Π_1). Since $\pi_1(s,t) = 0$ for either $s = 0$ or $t = 0$, we note $\Pi_1(X) = s_H(\Pi_1(X))s_K$ by Lemma 2.9. For a rank-one operator $X = \xi \otimes \eta^c$, (2.4) shows

$$\Pi_1(X) = \int_{-\infty}^{\infty} \left((s_H H)^{ix}\xi\right) \otimes \left((s_K K)^{ix}\eta\right)^c \frac{d\nu_1}{d|\nu_1|}(x)\, d|\nu_1|(x)$$

$$= \int_{-\infty}^{\infty} (s_H H)^{ix}(\xi \otimes \eta^c)(s_K K)^{-ix} d\nu_1(x)$$

$$= \int_{-\infty}^{\infty} (s_H H)^{ix} X (s_K K)^{-ix} d\nu_1(x),$$

which remains of course valid for finite-rank operators. Actually this integral expression for $\Pi_1(X)$ is also valid for an arbitrary operator $X \in B(\mathcal{H})$. In fact, as in the proof of Lemma 2.8 we approximate X by the sequence $\{p_n X p_n\}_{n=1,2,\ldots}$. At first $\Pi_1(p_n X p_n)$ tends to $\Pi_1(X)$ in the weak operator topology as remarked there. Therefore, it suffices to show the weak convergence

$$\int_{-\infty}^{\infty} (Hs_H)^{ix} p_n X p_n (Ks_K)^{-ix} d\nu_1(x) \longrightarrow \int_{-\infty}^{\infty} (Hs_H)^{ix} X (Ks_K)^{-ix} d\nu_1(x).$$

However, it simply follows from the Lebesgue dominated convergence theorem.

We next consider π_2 (with the double integral transformation Π_2). By Lemma 2.9 (and Remark 2.5) we have

$$\Pi_2(X) = s_H(\Pi_2(X))(1 - s_K) = \pi_2(H,0)s_H X(1 - s_K).$$

Recall $\pi_2(s,0) = \hat{\nu}_2(\log s)$ $(s > 0)$ so that

$$\pi_2(H,0)s_H = \int_{(0,\|H\|]} \hat{\nu}_2(\log s)\, dE_s$$

$$= \int_{(0,\|H\|]} \left(\int_{-\infty}^{\infty} s^{ix} d\nu_2(x)\right) dE_s = \int_{-\infty}^{\infty} (Hs_H)^{ix} d\nu_2(x)$$

due to the Fubini theorem. Therefore, we have

$$\Pi_2(X) = \int_{-\infty}^{\infty} (Hs_H)^{ix} X(1 - s_K)\, d\nu_2(x).$$

Symmetric arguments also show

$$\Pi_3(X) = (1 - s_H)X s_K \pi_3(0,K) = \int_{-\infty}^{\infty} (1 - s_H)X(Ks_K)^{-ix} d\nu_3(x)$$

while

$$\Pi_4(X) = (1 - s_H)(\Pi_4(X))(1 - s_K) = a(1 - s_H)X(1 - s_K)$$

is just trivial. By summing up all the Π_k's computed so far, we get the desired expression for $\Pi(X)$. □

Proposition 2.11. *Let $\pi(s,t)$ be the Schur multiplier in the previous lemma. If $\phi(s,t)$ is a Schur multiplier relative to a pair (H,K), then so is the pointwise product $\psi(s,t) = \pi(s,t)\phi(s,t)$. Furthermore, for each $X \in B(\mathcal{H})$ the corresponding double integral transformations $\Phi(X)$ and $\Psi(X)$ are related by*

$$\Psi(X) = \int_{-\infty}^{\infty} (H s_H)^{ix}(\Phi(X))(K s_K)^{-ix} d\nu_1(x)$$

$$+ \int_{-\infty}^{\infty} (H s_H)^{ix}(\Phi(X))(1 - s_K) d\nu_2(x)$$

$$+ \int_{-\infty}^{\infty} (1 - s_H)(\Phi(X))(K s_K)^{-ix} d\nu_3(x)$$

$$+a(1 - s_H)(\Phi(X))(1 - s_K).$$

Proof. The first statement follows from Lemmas 2.8 and 2.10. To get the expression for $\Psi(X)$, in the formula appearing in Lemma 2.10 we should just replace X by $\Phi(X)$. □

We end the chapter with the following remark on the standard 2×2-matrix trick, that will be sometimes useful in later chapters:

Remark 2.12. We set $\tilde{H} = \begin{bmatrix} H & 0 \\ 0 & K \end{bmatrix}$, and assume that ϕ is a Schur multiplier relative to $(\tilde{H}, \tilde{H})$ (or equivalently, so is ϕ relative to (H, H), (H, K) and (K, K)). Then, ϕ (on $[0, \|\tilde{H}\|] \times [0, \|\tilde{H}\|]$) admits an integral expression as (2.3) relative to $(\tilde{H}, \tilde{H})$. For $\tilde{X} = \begin{bmatrix} 0 & X \\ 0 & 0 \end{bmatrix}$ we compute

$$\alpha(\tilde{H},x)\tilde{X}\beta(\tilde{H},x) = \begin{bmatrix} \alpha(H,x) & 0 \\ 0 & \alpha(K,x) \end{bmatrix}\begin{bmatrix} 0 & X \\ 0 & 0 \end{bmatrix}\begin{bmatrix} \beta(H,x) & 0 \\ 0 & \beta(K,x) \end{bmatrix}$$

$$= \begin{bmatrix} 0 & \alpha(H,x)X\beta(K,x) \\ 0 & 0 \end{bmatrix}.$$

Therefore, the $(1,2)$-component of $\Phi_{\tilde{H},\tilde{H}}(\tilde{X}) = \int_{\Omega} \alpha(\tilde{H},x)\tilde{X}\beta(\tilde{H},x)\,d\sigma(x)$ is exactly

$$\int_{\Omega} \alpha(H,x)X\beta(K,x)\,d\sigma(x) = \Phi_{H,K}(X).$$

The support projection of $\tilde{H}$ is of the form

$$s_{\tilde{H}} = \begin{bmatrix} s_H & 0 \\ 0 & s_K \end{bmatrix},$$

and $s_H(\Phi_{H,K}(X))s_K$ is the $(1,2)$-component of $s_{\tilde{H}}\left(\Phi_{\tilde{H},\tilde{H}}(\tilde{X})\right)s_{\tilde{H}}$.

2.5 Notes and references

Motivated from perturbations of a continuous spectrum, scattering theory and triangular representations of Volterra operators (see [30]) as well as study of Hankel operators (see [71] for recent progress of the subject matter), in [14, 15, 16] M. Sh. Birman and M. Z. Solomyak systematically developed theory of double integral transformations formally written as

$$Y = \iint \phi(s,t)\,dF_t X\,dE_s$$

Besides the definition given at the beginning of this chapter (first defined on $\mathcal{C}_2(\mathcal{H})$), another definition by repeated integration

$$Y(s) = \left(\int \phi(s,t)\,dF_t \right) X, \quad Y = \int Y(s)\,dE_s \tag{2.24}$$

was also taken by Birman and Solomyak, where the latter integration is understood as the limit of Riemann-Stieltjes sums. Indeed, the articles [15, 16] were largely devoted to the well-definedness of the repeated integration in certain symmetric operator ideals in cases when ϕ is a function in some classes of Lipschitz type or of Sobolev type. For example, the following criterion was obtained:

Theorem Let $\phi(s,t)$ be a bounded Borel function on $[a,b] \times [c,d]$ satisfying Lip α with respect to variable s with a constant (of Hölder continuity of order α) independent of t. Assume that E_s and F_t are supported in $[a,b]$ and $[c,d]$ respectively. If $\alpha > \frac{1}{2}$, then ϕ is a Schur multiplier and for any $X \in B(\mathcal{H})$ the repeated integral (2.24) exists and coincides with $\Phi(X)$ (defined in §2.1). If $\alpha \leq \frac{1}{2}$, then for any $X \in \mathcal{C}_p(\mathcal{H})$ with $\frac{1}{p} > \frac{1}{2} - \alpha$ the repeated integral (2.24) exists as a compact operator.

But this type of results are not so useful in the present monograph because we mostly treat means (introduced in Definition 3.1) which do not at all satisfy the Lipschitz type condition.

As was shown in [69, 70] (also [15]), double integral transformations are closely related to problems of operator perturbations. For a C^1-function φ on an interval I ($\subseteq \mathbf{R}$) and self-adjoint operators $A = \int s\,dE_s$, $B = \int t\,dF_t$ with spectra contained in I we formally have

$$\varphi(A) - \varphi(B) = \int_I \int_I \varphi^{[1]}(s,t)\,dE_s(A-B)dF_t \tag{2.25}$$

with the divided difference

$$\varphi^{[1]}(s,t) = \begin{cases} \dfrac{\varphi(s) - \varphi(t)}{s-t} & \text{(if } s \neq t), \\[2ex] \varphi'(s) & \text{(if } s = t). \end{cases}$$

If $\varphi^{[1]}(s,t)$ is known to be a Schur multiplier relative to say some p-Schatten ideal $\mathcal{C}_p(\mathcal{H})$, then (2.25) for $A - B$ sitting in the ideal is justified and hence one gets the perturbation norm inequality

$$\|\varphi(A) - \varphi(B)\|_p \leq \text{const.} \|A - B\|_p, \qquad (2.26)$$

showing $\varphi(A) - \varphi(B) \in \mathcal{C}_p(\mathcal{H})$, i.e., the stability of perturbation. The following is a folk result (whose proof is an easy but amusing exercise): If $\varphi(s)$ is of the form $\varphi(s) = \int_{-\infty}^{\infty} e^{ist} d\nu(t)$ with a signed measure ν satisfying $\int_{-\infty}^{\infty} (1 + |t|) d|\nu|(t) < \infty$, then $\varphi^{[1]}(s,t)$ is a Schur multiplier relative to $\mathcal{C}_1(\mathcal{H})$ (and hence relative to any $\mathcal{C}_p(\mathcal{H})$). On the other hand, in [27] Yu. B. Farforovskaya obtained an example of $\varphi \in C^1(I)$ for which (2.26) fails to hold for $\|\cdot\|_1$. The next result due to E. B. Davies is very powerful:

Theorem ([24, Theorem 17]) Let φ be a function of the form

$$\varphi(s) = as + b + \int_{-\infty}^{s} (s - t) d\nu(t)$$

with $a, b \in \mathbf{R}$ and a signed measure ν of compact support. Then, the estimate (2.26) is valid for any $p \in (1, \infty)$.

The following "unitary version" of (2.25) is also useful: If φ is a C^1-function on the unit circle $\mathbf{T}$ (with a Schur multiplier $\varphi^{[1]}(s,t)$), then we have

$$\varphi(U) - \varphi(V) = \int_{\mathbf{T}} \int_{\mathbf{T}} \varphi^{[1]}(\zeta, \eta) \, dE_\zeta (U - V) \, dF_\eta$$

for unitary operators $U = \int_{\mathbf{T}} \zeta \, dE_\zeta$, $V = \int_{\mathbf{T}} \eta \, dF_\eta$. This technique was often used in M. G. Krein's works and is closely related to his famous spectral shift function.

Peller's characterization theorem (Theorem 2.2) was given in [69] ([70] is an announcement) while general results such as Propositions 2.6 and 2.7 were shown in [15, 16] by M. Sh. Birman and M. Z. Solomyak. Unfortunately these articles [15, 16, 69] (especially [69]) were not widely circulated. Our arguments here are basically taken from their articles, but we have tried to present more details. In fact, for the reader's convenience we have supplied some arguments that were omitted in the original articles.

3

Means of operators and their comparison

From now on we will study means $M(H,K)X$ of operators H, K, X with $H, K \geq 0$ (for certain scalar means $M(s,t)$). In fact, our operator means $M(H,K)X$ are defined as double integral transformations studied in Chapter 2 so that corresponding scalar means $M(s,t)$ are required to be Schur multipliers. In this chapter general properties of such operator means are clarified while some special series of concrete means will be exemplified in later chapters. Here we are mostly concerned with integral expressions (Theorem 3.4), comparison of norms (Theorem 3.7), norm estimate (Theorem 3.12) and the determination of the kernel and the closure of the range of the "mean transform" $M(H,K)$ (Theorem 3.16).

3.1 Symmetric homogeneous means

We begin by introducing a class of means for positive scalars and a partial order among them. This order will be quite essential in the sequel of the monograph. We confine ourselves to that class of means for convenience sake while all the results in the next §3.2 remain valid (with obvious modification) for more general means (as will be briefly discussed in §A.1).

Definition 3.1. A continuous positive real function $M(s,t)$ for $s, t > 0$ is called a *symmetric homogeneous mean* (or simply a *mean*) if M satisfies the following properties:

(a) $M(s,t) = M(t,s)$,
(b) $M(rs, rt) = rM(s,t)$ for $r > 0$,
(c) $M(s,t)$ is non-decreasing in s, t,
(d) $\min\{s,t\} \leq M(s,t) \leq \max\{s,t\}$.

We denote by $\mathfrak{M}$ the set of all such symmetric homogeneous means.

Definition 3.2. We assume $M, N \in \mathfrak{M}$. We write $M \preceq N$ when the ratio $M(e^x, 1)/N(e^x, 1)$ is a positive definite function on $\mathbf{R}$, or equivalently, the

matrix $\left[\dfrac{M(s_i, s_j)}{N(s_i, s_j)}\right]_{i,j=1,\cdots,n}$ is positive semi-definite for any $s_1, \ldots, s_n > 0$
with any size n. By the Bochner theorem it is also equivalent to the existence
of a symmetric probability measure ν on $\mathbf{R}$ satisfying $M(e^x, 1) = \hat{\nu}(x)N(e^x, 1)$
$(x \in \mathbf{R})$, that is,

$$M(s,t) = \hat{\nu}(\log s - \log t)N(s,t) \qquad (s,t > 0). \tag{3.1}$$

Here, $\hat{\nu}(x)$ means the Fourier transform $\hat{\nu}(x) = \displaystyle\int_{-\infty}^{\infty} e^{ixy} d\nu(y) \ (x \in \mathbf{R})$.

The reason why a symmetric probability ν comes out is that the real function $M(e^x, 1)/N(e^x, 1)$ takes value 1 at the origin. (See [39, Theorem 1.1] for
details.) Also, note that the order $M \preceq N$ is strictly stronger than the usual
(point-wise) order $M(s,t) \leq N(s,t)$ $(s,t > 0)$ (see [39, Example 3.5]). The
domain of $M \in \mathfrak{M}$ naturally extends to $[0, \infty) \times [0, \infty)$ in the following way:

$$M(s,0) = \lim_{t \searrow 0} M(s,t) \ (s > 0), \quad M(0,t) = \lim_{s \searrow 0} M(s,t) \ (t > 0),$$

$$M(0,0) = \lim_{s \searrow 0} M(s,0) = \lim_{t \searrow 0} M(0,t),$$

and $M(s,t)$ remains continuous on the extended domain. It is easy to check

$$M(s,0) = M(0,s) = sM(1,0) \qquad (s > 0) \tag{3.2}$$

and hence

$$M(0,0) = 0. \tag{3.3}$$

The most familiar means in $\mathfrak{M}$ are probably

$$A(s,t) = \frac{s+t}{2} \qquad\qquad\qquad \text{(arithmetic mean)},$$

$$L(s,t) = \frac{s-t}{\log s - \log t} = \int_0^1 s^x t^{1-x} dx \quad \text{(logarithmic mean)},$$

$$G(s,t) = \sqrt{st} \qquad\qquad\qquad \text{(geometric mean)},$$

$$M_{\text{har}}(s,t) = \frac{2}{s^{-1} + t^{-1}} \qquad\qquad \text{(harmonic mean)}.$$

The largest and smallest means in $\mathfrak{M}$

$$M_\infty(s,t) = \max\{s,t\} \quad \text{and} \quad M_{-\infty}(s,t) = \min\{s,t\}$$

will play an important role in our discussions below.

We have the following order relation among the above means:

$$M_{-\infty} \preceq M_{\text{har}} \preceq G \preceq L \preceq A \preceq M_\infty. \tag{3.4}$$

The proof is found in the more general [39, Theorem 2.1] (i.e., (5.2) right
before Theorem 5.1 in Chapter 5; see also [38, Proposition 1 or more generally

Theorem 5]). However, here for the reader's convenience we prove this special case by bare-handed computations. Firstly Example 3.6, (c) below shows $A \preceq M_\infty$. For $L \preceq A$ we just note

$$\frac{L(e^x, 1)}{A(e^x, 1)} = \frac{e^x - 1}{x} \times \frac{2}{e^x + 1} = \frac{2 \sinh(x/2)}{x \cosh(x/2)} = \int_0^1 \frac{\cosh(ax/2)}{\cosh(x/2)} \, da.$$

Since $\cosh(ax/2)/\cosh(x/2)$ is positive definite for each $a \in [0,1]$ (see §6.3, 1), so is the above integral. (The Fourier transform can be also explicitly determined; see (6.8) or the computations in [38, p. 305].) For $G \preceq L$ we observe

$$\frac{G(e^x, 1)}{L(e^x, 1)} = e^{x/2} \times \frac{x}{e^x - 1} = \frac{x}{2 \sinh(x/2)}.$$

The well-known formula

$$\int_{-\infty}^{\infty} \frac{x}{2 \sinh(x/2)} e^{ixy} dx = \frac{1}{4 \cosh^2(\pi y)} \tag{3.5}$$

and its inverse transform guarantee the positive definiteness of the ratio. Finally, both of

$$\frac{M_{\mathrm{har}}(e^x, 1)}{G(e^x, 1)} = \frac{2}{e^{-x} + 1} \times e^{-x/2} = \frac{1}{\cosh(x/2)},$$

$$\frac{M_{-\infty}(e^x, 1)}{M_{\mathrm{har}}(e^x, 1)} = \min\{e^x, 1\} \times \frac{e^{-x} + 1}{2} = \frac{e^{-|x|} + 1}{2}$$

are obviously positive definite (see (5.8) and (7.3)), and we are done.

Now let H, K be positive operators in $B(\mathcal{H})$ with the spectral decompositions $H = \int_0^{\|H\|} s \, dE_s$ and $K = \int_0^{\|K\|} t \, dF_t$. For a mean $M \in \mathfrak{M}$ we would like to define the corresponding double integral transformation relative to the pair (H, K):

$$M(H, K)X = M_{H,K}(X) = \int_0^{\|H\|} \int_0^{\|K\|} M(s, t) \, dE_s X dF_t$$

for $X \in B(\mathcal{H})$, and we consider this transformation acting on operators on $\mathcal{H}$ as a "mean of H and K". The transformation $M(H, K)$ always makes sense if restricted on the Hilbert-Schmidt class $\mathcal{C}_2(\mathcal{H})$ (in particular, on the ideal $\mathcal{I}_{\mathrm{fin}}$); it is the function calculus on $\mathcal{C}_2(\mathcal{H})$ via $M(s, t)$ of the left multiplication by H and the right multiplication by K. But, to define $M(H, K) = M_{H,K}$ on the whole $B(\mathcal{H})$, we have to verify that M is a Schur multiplier relative to (H, K). For instance, if H, K have finite spectra so that they have the discrete spectral decompositions

$$H = \sum_{i=1}^m s_i P_i \quad \text{and} \quad K = \sum_{j=1}^n t_j Q_j$$

with projections P_i, Q_j such that $\sum_{i=1}^{m} P_i = \sum_{j=1}^{n} Q_j = 1$, then each $M \in \mathfrak{M}$ is a Schur multiplier relative to (H, K) and

$$M(H, K)X = \sum_{i=1}^{m} \sum_{j=1}^{n} M(s_i, t_j) P_i X Q_j$$

(this is the case even for any Borel function on $[0, \infty) \times [0, \infty)$).

In what follows we simply say that $M \in \mathfrak{M}$ is a *Schur multiplier* if it is so relative to any pair (H, K) of positive operators. As for the means A, L and G, the corresponding double integral transformations have the concrete forms

$$A(H, K)X = \frac{1}{2}(HX + XK),$$

$$L(H, K)X = \int_0^1 H^x X K^{1-x} dx,$$

$$G(H, K)X = H^{\frac{1}{2}} X K^{\frac{1}{2}},$$

showing that they are indeed Schur multipliers. But it is not so obvious to determine whether a given $M \in \mathfrak{M}$ is a Schur multiplier. The next proposition provides a handy sufficient condition.

Proposition 3.3. *Let* $M, N \in \mathfrak{M}$ *and* H, K *be positive operators.*

(a) *If* $M \preceq N$ *and* N *is a Schur multiplier relative to* (H, K), *then so is* M.
(b) *If* $M \preceq M_\infty$, *then* M *is a Schur multiplier* (*relative to any* (H, K)).

Proof. (a) By Definition 3.2 there exists a symmetric probability measure ν satisfying (3.1). Noting $M(1, 0) \leq N(1, 0)$ (following from $M(s, t) \leq N(s, t)$ when $s, t > 0$) we set $c = M(1, 0)/N(1, 0)$ if $N(1, 0) > 0$, otherwise $c = 0$. Then, thanks to (3.2) and (3.3) we have $M(s, t) = \pi(s, t) N(s, t)$ for all $s, t \geq 0$ with

$$\pi(s, t) = \begin{cases} \hat{\nu}(\log s - \log t) & \text{if } s, t > 0, \\ c & \text{if } s > 0 \text{ and } t = 0, \\ c & \text{if } s = 0 \text{ and } t > 0, \\ 0 & \text{if } s = t = 0. \end{cases} \tag{3.6}$$

Hence the assertion is a consequence of Proposition 2.11 (based on Lemma 2.8).

(b) By virtue of (a) it suffices to show that M_∞ is a Schur multiplier. Since A is obviously a Schur multiplier (as was mentioned above), (a) implies by (3.4) that $M_{-\infty}$ is a Schur multiplier. Hence so is M_∞ because of the simple formula

$$M_\infty(s, t) = 2A(s, t) - M_{-\infty}(s, t) \tag{3.7}$$

(see the remark after Lemma 2.8). □

The fact that $M_{\pm\infty}$ are Schur multipliers can be also seen from the discrete decompositions explained in §A.3 (see (A.4) and Theorem A.6). All the concrete means treated in later chapters satisfy $M \preceq M_\infty$ so that they are all Schur multipliers. It is easy to write down examples of $M \in \mathfrak{M}$ not satisfying $M \preceq M_\infty$; nevertheless we have so far no explicit example of $M \in \mathfrak{M}$ which is not a Schur multiplier.

3.2 Integral expression and comparison of norms

We begin with the integral expression (Theorem 3.4) for operator means, which is an adaptation of the integral expression in Proposition 2.11 (also Lemma 2.9) in the present setting of means in $\mathfrak{M}$. (Similar integral expressions for wider classes of means will be worked out in §8.1 and §A.1.) Then, comparison of norms of means will be an easy consequence.

In [49, p. 138] the following formula appears as an exercise:

$$\sum_{t\in\mathbf{R}} |\mu(\{t\})|^2 = \lim_{T\to\infty} \frac{1}{2T} \int_{-T}^{T} |\hat{\mu}(t)|^2 dt$$

for a complex measure μ on $\mathbf{R}$. A related fact will be needed in the proof of the theorem, and the proofs for this fact as well as the above formula will be presented in §A.4 for the reader's convenience.

Theorem 3.4. *Let $M, N \in \mathfrak{M}$ and H, K be positive operators. If $M \preceq N$ with the representing measure ν for $M(e^x, 1)/N(e^x, 1)$ (see Definition 3.2) and if N is a Schur multiplier relative to (H, K), then so is M and*

$$M(H,K)X = \int_{-\infty}^{\infty} (Hs_H)^{ix}(N(H,K)X)(Ks_K)^{-ix} d\nu(x)$$
$$+ M(1,0)(HX(1-s_K) + (1-s_H)XK) \qquad (3.8)$$

for all $X \in B(\mathcal{H})$. In this case we also have

$$M(H,K)X = \int_{\{x\neq 0\}} (Hs_H)^{ix}(N(H,K)X)(Ks_K)^{-ix} d\nu(x)$$
$$+ \nu(\{0\})N(H,K)X. \qquad (3.9)$$

Proof. We use the same notations as in the proof of Proposition 3.3, (a). Use of Lemma 2.9 (see (2.23)) to N with (3.2) and (3.3) yields

$$N(H,K)X = s_H(N(H,K)X)s_K$$
$$+ N(1,0)(HX(1-s_K) + (1-s_H)XK). \qquad (3.10)$$

Since $M(s,t) = \pi(s,t)N(s,t)$ for all $s,t \geq 0$ with π defined by (3.6), Proposition 2.11 implies

$$M(H,K)X = \int_{-\infty}^{\infty} (Hs_H)^{ix}(N(H,K)X)(Ks_K)^{-ix}d\nu(x)$$
$$+ c\left(s_H(N(H,K)X)(1-s_K) + (1-s_H)(N(H,K)X)s_K\right).$$

Since $M(1,0) = cN(1,0)$, the expression (3.8) is obtained by substituting (3.10) into the above integral expression.

To show (3.9), we begin with the claim $M(1,0) = \nu(\{0\})N(1,0)$. When $N(1,0) = 0$, we must have $M(1,0) = 0$ due to $M(1,0) \le N(1,0)$ and there is nothing to prove. Thus we may and do assume $N(1,0) > 0$. In this case we note

$$\lim_{x\to-\infty} \hat{\nu}(x) = \lim_{x\to-\infty} \frac{M(e^x,1)}{N(e^x,1)} = \frac{M(0,1)}{N(0,1)} \left(= \frac{M(1,0)}{N(1,0)}\right),$$
$$\lim_{x\to\infty} \hat{\nu}(x) = \lim_{x\to\infty} \frac{M(e^x,1)}{N(e^x,1)} = \lim_{x\to\infty} \frac{M(1,e^{-x})}{N(1,e^{-x})} = \frac{M(1,0)}{N(1,0)}.$$

Therefore, we conclude

$$\lim_{x\to\pm\infty} \hat{\nu}(x) = \frac{M(1,0)}{N(1,0)},$$

and the claim follows from Corollary A.8 in §A.4. The claim and (3.8) yield

$$M(H,K)X$$
$$= \int_{\{x\neq0\}} (Hs_H)^{ix}(N(H,K)X)(Ks_K)^{-ix}d\nu(x)$$
$$+ \nu(\{0\})\left(s_H(N(H,K)X)s_K + N(1,0)(HX(1-s_K) + (1-s_H)XK)\right)$$
$$= \int_{\{x\neq0\}} (Hs_H)^{ix}(N(H,K)X)(Ks_K)^{-ix}d\nu(x) + \nu(\{0\})N(H,K)X.$$

Here, the second equality is due to (3.10). □

From the expression (3.9) in the preceding theorem and Theorem A.5 we have

Corollary 3.5. *Let $M, N \in \mathfrak{M}$ ($M \preceq N$) and H, K be as in the theorem. Then for any unitarily invariant norm $|||\cdot|||$ we have*

$$|||M(H,K)X||| \le |||N(H,K)X|||$$

for all $X \in B(\mathcal{H})$. In particular,

$$\|M(H,K)\|_{(|||\cdot|||,\,|||\cdot|||)} \le \|N(H,K)\|_{(|||\cdot|||,\,|||\cdot|||)}.$$

Example 3.6. The following examples are applications of the integral expression in the above theorem to means in (3.4).

(a) Since the ratio $G(e^x, 1)/A(e^x, 1) = \left(\cosh\left(\frac{x}{2}\right)\right)^{-1}$ is the Fourier transform of $\left(\cosh(\pi x)\right)^{-1}$,

$$H^{\frac{1}{2}}XK^{\frac{1}{2}} = \int_{-\infty}^{\infty} (H_{SH})^{ix}(HX + XK)(K_{SK})^{-ix} \frac{dx}{2\cosh(\pi x)}.$$

Actually, the observation of this expression is the starting point of our works on means of operators in a series of recent articles ([54, 38, 39]). We also point out that the use of this integral transformation was crucial in [22, 23].

(b) Since $M_{-\infty}(e^x, 1)/G(e^x, 1) = e^{-|x|/2}$ is the Fourier transform of $\frac{1}{2\pi}\left(x^2 + \frac{1}{4}\right)^{-1}$ (see (5.8) and (7.3)),

$$M_{-\infty}(H, K)X = \int_{-\infty}^{\infty} (H_{SH})^{\frac{1}{2}+ix}X(K_{SK})^{\frac{1}{2}-ix} \frac{dx}{2\pi\left(x^2 + \frac{1}{4}\right)}.$$

(c) Since $A(e^x, 1)/M_{\infty}(e^x, 1) = \frac{1}{2}(1 + e^{-|x|})$ is the Fourier transform of the measure $\frac{1}{2}\delta_0 + \frac{1}{2\pi}(x^2 + 1)^{-1}dx$,

$$HX + XK = M_{\infty}(H, K)X$$
$$+ \int_{-\infty}^{\infty} (H_{SH})^{ix}(M_{\infty}(H, K)X)(K_{SK})^{-ix} \frac{dx}{\pi(x^2 + 1)}.$$

The opposite direction of this is also possible. Since $M_{-\infty}(e^x, 1)/A(e^x, 1) = 2e^{-|x|}/(1 + e^{-|x|}) = e^{-|x|/2}/\cosh\left(\frac{x}{2}\right)$ is the Fourier transform of the convolution product

$$f(x) = \left(\frac{1}{\cosh(\pi x)}\right) * \left(\frac{1}{2\pi\left(x^2 + \frac{1}{4}\right)}\right),$$

one obtains thanks to (3.7)

$$M_{\infty}(H, K)X = HX + XK$$
$$- \frac{1}{2}\int_{-\infty}^{\infty} (H_{SH})^{ix}(HX + XK)(K_{SK})^{-ix}f(x)\,dx. \quad (3.11)$$

The general comparison theorem for means in $\mathfrak{M}$ was summarized in [39, Theorem 1.1] in the setting of matrices, and its extension to the operator setting was stated at the end of [39]. However, the statement there is quite rough and its sketch for the proof contains some inaccurate arguments. So, for completeness let us prove the next theorem in a precise form.

Theorem 3.7. *For $M, N \in \mathfrak{M}$ the following conditions are all equivalent:*

(i) *there exists a symmetric probability measure ν on $\mathbf{R}$ with the following property: if N is a Schur multiplier relative to (H, K) of non-singular positive operators, then so is M and*

$$M(H,K)X = \int_{-\infty}^{\infty} H^{ix}(N(H,K)X)K^{-ix}d\nu(x)$$

for all $X \in B(\mathcal{H})$;

(ii) *if N is a Schur multiplier relative to a pair (H,K) of positive operators, then so is M and*

$$|||M(H,K)X||| \leq |||N(H,K)X|||$$

for all unitarily invariant norms and all $X \in B(\mathcal{H})$;

(iii) $\|M(H,H)X\| \leq \|N(H,H)X\|$ *for all $H \geq 0$ and all $X \in \mathcal{I}_{fin}$;*

(iv) $M \preceq N$.

Proof. (iv) $\Rightarrow$ (i) is contained in Theorem 3.4, and (iv) $\Rightarrow$ (ii) follows from Corollary 3.5. When H, K and X are of finite-rank, (ii) and (iii) reduce to the same condition in the matrix case (of any size). So (ii) $\Rightarrow$ (iv) and (iii) $\Rightarrow$ (iv) are seen from [39, Theorem 1.1]. (Necessary arguments under a slightly weaker assumption will be actually presented in the proof of Theorem A.3 in §A.1.) For (i) $\Rightarrow$ (iv) put $H = s1$ $(s > 0)$ and $K = X = 1$; then the integral expression in (i) reduces to $M(s,1) = \hat{\nu}(\log s)N(s,1)$, i.e., $M \preceq N$.

It remains to show (iv) $\Rightarrow$ (iii), which is not quite trivial because N in (iii) is not a priori a Schur multiplier relative to (H,H). At first, when H is also of finite-rank, the inequality in (iii) follows from (iv) by [39, Theorem 1.1] (or from (ii) since we have already had (iv) $\Rightarrow$ (ii)). For a general H choose a sequence $\{H_n\}$ of finite-rank positive operators such that $\|H_n\| \leq \|H\|$ and $H_n \to H$ in the strong operator topology. Then $\pi_\ell(H_n) \to \pi_\ell(H)$ strongly on $\mathcal{C}_2(\mathcal{H})$ because for a rank-one operator $\xi \otimes \eta^c$ we get

$$\|\pi_\ell(H_n)(\xi \otimes \eta^c) - \pi_\ell(H)(\xi \otimes \eta^c)\|_2$$
$$= \|(H_n\xi - H\xi) \otimes \eta^c\|_2 = \|H_n\xi - H\xi\| \times \|\eta\| \longrightarrow 0.$$

Similarly $\pi_r(H_n) \to \pi_r(H)$ strongly on $\mathcal{C}_2(\mathcal{H})$. Since $M(s,t)$ is uniformly approximated on $[0, \|H\|] \times [0, \|H\|]$ by polynomials in two variables s and t, it follows that $M(H_n, H_n) \to M(H,H)$ strongly on $\mathcal{C}_2(\mathcal{H})$. For every $X \in \mathcal{I}_{fin}$ $(\subseteq \mathcal{C}_2(\mathcal{H}))$ we thus get

$$\|M(H_n, H_n)X - M(H,H)X\| \leq \|M(H_n, H_n)X - M(H,H)X\|_2 \to 0$$

and the same is true for N too. Hence the required inequality is obtained by taking the limit from $\|M(H_n, H_n)X\| \leq \|N(H_n, H_n)X\|$. □

3.3 Schur multipliers for matrices

In estimating the norm of a double integral transformation, it is sometimes useful to reduce the problem to the matrix case by approximation (though computing the Schur multiplication norm is usually difficult even in the matrix case). Such an approximation technique is developed here, which will be indispensable in §3.5.

We begin with basics on Schur multiplication on matrices. Let $A = [a_{ij}]_{i,j=1,2,\dots}$ be an infinite complex matrix such that $\sup_{i,j} |a_{ij}| < \infty$. Then one can formally define a *Schur multiplication operator* S_A on the space of infinite matrices as

$$S_A(X) = A \circ X = [a_{ij} x_{ij}] \quad \text{for } X = [x_{ij}],$$

where $\circ$ is the *Schur product* or the *Hadamard product* (i.e., the entry-wise product). Consider the Hilbert space ℓ^2 with the canonical basis $\{e_i\}_{i=1,2,\dots}$ and identify an operator $X \in B(\ell^2)$ as the matrix $\left[(Xe_j, e_i)\right]_{i,j=1,2,\dots}$. We then say that A is a *Schur multiplier* if S_A gives rise to a bounded transformation of $C_1(\ell^2)$ into itself (or equivalently, of $B(\ell^2)$ into itself). A Schur multiplication operator S_A as above is realized as a double integral transformation of discrete type. In fact, assume that $H, K \geq 0$ are diagonalizable with

$$H = \sum_{i=1}^{\infty} s_i \xi_i \otimes \xi_i^c \quad \text{and} \quad K = \sum_{i=1}^{\infty} t_i \eta_i \otimes \eta_i^c$$

for some orthonormal bases $\{\xi_i\}$ and $\{\eta_i\}$. For any Borel function ϕ on $[0, \infty) \times [0, \infty)$ the corresponding double integral transformation $\Phi_{H,K}$ can be represented as

$$\Phi_{H,K}(UXV^*) = US_A(X)V^* \quad \text{for } X = [x_{ij}] \in B(\ell^2), \qquad (3.12)$$

where $A = [\phi(s_i, t_j)]_{i,j=1,2,\dots}$ and U, V are unitary operators given by $Ue_i = \xi_i$, $Ve_i = \eta_i$. In this way, ϕ is a Schur multiplier relative to (H, K) if and only if $A = [\phi(s_i, t_j)]$ is a Schur multiplier, and in this case

$$\|\Phi_{H,K}\|_{(1,1)} = \|S_A\|_{(1,1)}. \qquad (3.13)$$

Moreover, the characterization (iv) of Theorem 2.2 reads as follows: there exist a Hilbert space $\mathcal{K}$ ($= L^2(\Omega, \sigma)$ there) and bounded sequences $\{u_i\}$ and $\{v_j\}$ of vectors in $\mathcal{K}$ such that

$$a_{ij} \,(= \phi(s_i, t_j)) = (u_i, v_j)_{\mathcal{K}} \qquad (i, j = 1, 2, \dots).$$

This criterion (known as Haagerup's criterion) was independently obtained by U. Haagerup (see **4** in §3.7). In particular, when $A = [a_{ij}]_{i,j=1,\dots,n}$ is an $n \times n$ matrix, the Schur multiplication operator S_A is defined on $M_n(\mathbf{C})$, the algebra of $n \times n$ matrices, and furthermore the following is known (see **4** in §3.7):

$$\|S_A\|_{(1,1)} \left(= \|S_A\|_{(\infty,\infty)}\right)$$
$$= \min\{\kappa \geq 0 : \text{there are } \xi_1, \dots, \xi_n, \eta_1, \dots, \eta_n \in \mathbf{C}^n \text{ such that}$$
$$\|\xi_i\| \leq \kappa^{1/2}, \|\eta_j\| \leq \kappa^{1/2}, a_{ij} = (\xi_i, \eta_j) \text{ for } i, j = 1, \dots, n\}. \quad (3.14)$$

If A is a positive semi-definite matrix, then

$$\|S_A\|_{(\infty,\infty)} = \max_i a_{ii}. \tag{3.15}$$

In fact, this is immediately seen from (3.14); if $\xi_1, \ldots, \xi_n$ are the row vectors of $A^{1/2}$, then $a_{ij} = (\xi_i, \xi_j)$ for all i, j. (A different proof without using (3.14) can be found in [4, 42].)

Lemma 3.8. *An infinite matrix $A = [a_{ij}]_{i,j=1,2,\ldots}$ is a Schur multiplier if and only if*

$$\sup_{n \geq 1} \left\| S_{[a_{ij}]_{i,j=1,\cdots,n}} \right\|_{(1,1)} < \infty.$$

In this case, $\|S_A\|_{(1,1)}$ is equal to the above supremum.

Proof. If A is a Schur multiplier, then it is obvious that

$$\left\| S_{[a_{ij}]_{i,j=1,\cdots,n}} \right\|_{(1,1)} \leq \|S_A\|_{(1,1)} \quad (\text{for } n = 1, 2, \ldots).$$

Conversely, assume that

$$\kappa = \sup_{n \geq 1} \left\| S_{[a_{ij}]_{i,j=1,\cdots,n}} \right\|_{(1,1)} < \infty.$$

Let $p_n = \sum_{i=1}^n e_i \otimes e_i^c$ with the canonical basis $\{e_i\}$ for ℓ^2. For every $X \in \mathcal{C}_1(\ell^2)$ and $n = 1, 2, \ldots$ we get

$$\|S_A(p_n X p_n)\|_1 = \|[a_{ij} x_{ij}]_{i,j=1,\cdots,n}\|_1 \leq \kappa \|p_n X p_n\|_1 \leq \kappa \|X\|_1,$$

and

$$\|S_A(p_m X p_m) - S_A(p_n X p_n)\|_1 \leq \kappa \|p_m X p_m - p_n X p_n\|_1.$$

By approximating X by finite-rank operators in the norm $\|\cdot\|_1$, one observes $\lim_{m,n\to\infty} \|p_m X p_m - p_n X p_n\|_1 = 0$ so that $\{S_A(p_n X p_n)\}_{n=1,2,\ldots}$ is Cauchy in $\mathcal{C}_1(\ell^2)$ from the second inequality and $\|S_A(p_n X p_n) - Y\|_1 \to 0$ for some $Y \in \mathcal{C}_1(\ell^2)$. Since the convergence also takes place in the weak operator topology, this limit Y must be equal to $S_A(X)$ and consequently

$$\|S_A(X)\|_1 = \lim_{n\to\infty} \|S_A(p_n X p_n)\|_1 \leq \kappa \|X\|_1$$

from the above first estimate. $\square$

The next lemma will play a key role in §3.5. The assumption of ϕ here may not be best possible, however it is enough for our purpose.

Lemma 3.9. *Let $\phi(s,t)$ be a function on $[0,\alpha] \times [0,\alpha]$ where $0 < \alpha < \infty$, and assume that ϕ is bounded and continuous at any point possibly except at $(0,0)$. Then the following conditions are equivalent:*

(i) *ϕ is a Schur multiplier relative to every pair (H, K) of positive operators with $\|H\|, \|K\| \leq \alpha$;*

(ii) $\sup\{\|S_{[\phi(s_i,s_j)]_{i,j=1,\cdots,n}}\|_{(1,1)} : 0 \le s_1,\ldots,s_n \le \alpha, n \ge 1\} < \infty$, where repetition is allowed for $s_1,\ldots,s_n$.

Furthermore, if (ii) holds with finite supremum κ, then $\|\Phi_{H,K}\|_{(1,1)} \le \kappa$ for any (H,K) with $\|H\|,\|K\| \le \alpha$.

Proof. (i) $\Rightarrow$ (ii). By assuming (i) and the failure of (ii), we will obtain a contradiction. Since (ii) fails to hold, for each n one can choose $s_1^{(n)},\ldots,s_n^{(n)}$ from $[0,\alpha]$ in such a way that

$$\sup_{n \ge 1} \left\|S_{\left[\phi(s_i^{(n)},s_j^{(n)})\right]_{i,j=1,\cdots,n}}\right\|_{(1,1)} = \infty.$$

Let $\{s_i\}_{i=1,2,\ldots}$ be the sequence

$$s_1^{(1)}, s_1^{(2)}, s_2^{(2)}, s_1^{(3)}, s_2^{(3)}, s_3^{(3)}, \cdots, s_1^{(n)},\ldots,s_n^{(n)},\cdots$$

obtained so far. We set $A = [\phi(s_i,s_j)]_{i,j=1,2,\ldots}$ and $H = \sum_{i=1}^{\infty} s_i \xi_i \otimes \xi_i^c$ where $\{\xi_i\}$ is an orthonormal basis. Then (i) implies that ϕ is a Schur multiplier relative to (H,H), so A must be a Schur multiplier as remarked just after (3.12). But, since $[\phi(s_i^{(n)},s_j^{(n)})]_{i,j=1,\cdots,n}$ is a principal submatrix of A, it is obvious that

$$\left\|S_{\left[\phi(s_i^{(n)},s_j^{(n)})\right]_{i,j=1,\cdots,n}}\right\|_{(1,1)} \le \|S_A\|_{(1,1)}$$

for all n. The supremum of the above left-hand side is ∞, a contradiction.

(ii) $\Rightarrow$ (i). Assume that the supremum κ in (ii) is finite. Let H be a positive operator with $\|H\| \le \alpha$ and the spectral decomposition $H = \int_0^{\alpha} s\,dE_s$. For each $n = 1,2,\ldots$ we divide $[0,\alpha]$ into subintervals

$$\Lambda_i^{(n)} = \left[\frac{i-1}{n}\alpha, \frac{i}{n}\alpha\right) \quad (i = 1,\ldots,n-1) \quad \text{and} \quad \Lambda_n^{(n)} = \left[\frac{n-1}{n}\alpha, \alpha\right],$$

and let $t_i^{(n)} = \frac{i-1}{n}\alpha$ $(i = 1,\ldots,n)$. Define

$$\phi_n(s,t) = \sum_{i,j=1}^n \phi(t_i^{(n)},t_j^{(n)})\chi_{\Lambda_i^{(n)} \times \Lambda_j^{(n)}}(s,t) \quad \text{for } (s,t) \in [0,\alpha] \times [0,\alpha]$$

and

$$H_n = \sum_{i=1}^n t_i^{(n)} E_{\Lambda_i^{(n)}}.$$

Then the double integral transformation $\Phi_n = \Phi_{H_n,H_n}$ corresponding to ϕ_n is given by

$$\Phi_n(X) = \sum_{i,j=1}^n \phi(t_i^{(n)},t_j^{(n)})E_{\Lambda_i^{(n)}} X E_{\Lambda_j^{(n)}}.$$

Since H_n is obviously diagonalizable, we write $H_n = \sum_{i=1}^{\infty} s_i^{(n)} \xi_i^{(n)} \otimes \xi_i^{(n)c}$ with an orthonormal basis $\{\xi_i^{(n)}\}_{i=1,2,\cdots}$ and set $A_n = \left[\phi(s_i^{(n)}, s_j^{(n)})\right]_{i,j=1,2,\cdots}$. Then, thanks to (3.13) we get

$$\|\Phi_n\|_{(1,1)} = \|S_{A_n}\|_{(1,1)}.$$

By assumption (ii) we apply Lemma 3.8 to conclude $\|S_{A_n}\|_{(1,1)} \leq \kappa$ so that $\|\Phi_n\|_{(1,1)} \leq \kappa$ for all n.

Now let $\xi, \eta, \xi', \eta' \in \mathcal{H}$ be arbitrary. For $\Phi = \Phi_{H,H}$ we get

$$
\begin{aligned}
\left(\Phi(\xi \otimes \eta^c)\xi', \eta'\right) &= \left(\Phi(\xi \otimes \eta^c), \eta' \otimes \xi'^c\right)_{C_2(\mathcal{H})} \\
&= \int_0^\alpha \int_0^\alpha \phi(s,t)\, d\left(E_s(\xi \otimes \eta^c)E_t, \eta' \otimes \xi'^c\right)_{C_2(\mathcal{H})} \\
&= \int_0^\alpha \int_0^\alpha \phi(s,t)\, d(E_s\xi, \eta')\, d(\xi', E_t\eta),
\end{aligned}
$$

and similarly

$$
\left(\Phi_n(\xi \otimes \eta^c)\xi', \eta'\right) = \int_0^\alpha \int_0^\alpha \phi_n(s,t)\, d(E_s\xi, \eta')\, d(\xi', E_t\eta).
$$

Here, the complex-valued measures $d(E_s\xi, \eta'), d(\xi', E_t\eta)$ are denoted by λ, μ respectively with their absolute values $|\lambda|, |\mu|$. By assumption, $|\phi(s,t)| \leq m$ (so $|\phi_n(s,t)| \leq m$ as well) on $[0,\alpha] \times [0,\alpha]$ for some $m < \infty$. For each $0 < \delta < \alpha$, since $\phi_n(0,0) = \phi(0,0)$, we estimate

$$
\begin{aligned}
&\left|\left(\Phi_n(\xi \otimes \eta^c)\xi', \eta'\right) - \left(\Phi(\xi \otimes \eta^c)\xi', \eta'\right)\right| \\
&\leq \left| \int_{([0,\alpha]\times[0,\alpha])\setminus([0,\delta)\times[0,\delta))} (\phi_n(s,t) - \phi(s,t))\, d(\lambda \times \mu)(s,t) \right| \\
&\quad + \left| \int_{([0,\delta)\times[0,\delta))\setminus\{(0,0)\}} \phi_n(s,t)\, d(\lambda \times \mu)(s,t) \right| \\
&\quad + \left| \int_{([0,\delta)\times[0,\delta))\setminus\{(0,0)\}} \phi(s,t)\, d(\lambda \times \mu)(s,t) \right| \\
&\leq \int_{([0,\alpha]\times[0,\alpha])\setminus([0,\delta)\times[0,\delta))} |\phi_n(s,t) - \phi(s,t)|\, d(|\lambda| \times |\mu|)(s,t) \\
&\quad + 2m(|\lambda| \times |\mu|)\left(([0,\delta)\times[0,\delta))\setminus\{(0,0)\}\right).
\end{aligned}
$$

For any $\delta > 0$ the first term of the latter expression tends to 0 as $n \to \infty$ because ϕ is continuous (hence uniformly continuous) on $([0,\alpha] \times [0,\alpha]) \setminus ([0,\delta) \times [0,\delta))$ so that $\phi_n \to \phi$ uniformly there. But the second term can be arbitrarily small when $\delta > 0$ is small enough. Therefore, we arrive at

$$
\lim_{n \to \infty} \left(\Phi_n(\xi \otimes \eta^c)\xi', \eta'\right) = \left(\Phi(\xi \otimes \eta^c)\xi', \eta'\right).
$$

This implies that $\Phi_n(X) \to \Phi(X)$ in the weak operator topology for all $X \in \mathcal{I}_{\text{fin}}$. Since $\|\Phi_n\|_{(1,1)} \le \kappa$ for all n as stated above, the lower semi-continuity of $\|\cdot\|_1$ in the weak operator topology (see [37, Proposition 2.11]) yields

$$\|\Phi(X)\|_1 \le \liminf_{n\to\infty} \|\Phi_n(X)\|_1 \le \kappa \|X\|_1$$

for all $X \in \mathcal{I}_{\text{fin}}$. For each $X \in \mathcal{C}_1(\mathcal{H})$ we approximate X by $p_n X p_n$ with finite-rank projections $p_n \nearrow 1$. Then $\{\Phi(p_n X p_n)\}$ is $\|\cdot\|_1$-Cauchy and $\Phi(p_n X p_n) \to Y \in \mathcal{C}_1(\mathcal{H})$ in the norm $\|\cdot\|_1$ as in the proof of Lemma 3.8. However, we claim $Y = \Phi(X)$. In fact, since Φ is a bounded operator on $\mathcal{C}_2(\mathcal{H})$, we have $\|\Phi(p_n X p_n) - \Phi(X)\|_2 \to 0$ (as well as $\|\Phi(p_n X p_n) - Y\|_2 \to 0$ thanks to $\|\cdot\|_2 \le \|\cdot\|_1$). Since $Y = \Phi(X)$, from the above estimate for operators in $\in \mathcal{I}_{\text{fin}}$ we have

$$\|\Phi(X)\|_1 = \lim_{n\to\infty} \|\Phi(p_n X p_n)\|_1 \le \kappa \|X\|_1$$

for all $X \in \mathcal{C}_1(\mathcal{H})$.

Finally, the standard 2×2-matrix trick can be conveniently used to extend this inequality to a pair (H, K) with $\|H\|, \|K\| \le \alpha$. In fact, with $\tilde{H}$ and $\tilde{X}$ as in Remark 2.12 we notice

$$\Phi_{\tilde{H}, \tilde{H}}(\tilde{X}) = \begin{bmatrix} 0 & \Phi_{H,K}(X) \\ 0 & 0 \end{bmatrix},$$

which implies

$$\|\Phi_{H,K}(X)\|_1 = \|\Phi_{\tilde{H}, \tilde{H}}(\tilde{X})\|_1 \le \kappa \|\tilde{X}\|_1 = \kappa \|X\|_1$$

for $X \in \mathcal{C}_1(\mathcal{H})$. Thus, ϕ is a Schur multiplier relative to (H, K) and $\|\Phi_{H,K}\|_{(1,1)} \le \kappa$. $\square$

3.4 Positive definite kernels

We say that $M \in \mathfrak{M}$ is a *positive definite kernel* if $[M(s_i, s_j)]_{i,j=1,\cdots,n}$ is positive semi-definite for any $s_1, \ldots, s_n > 0$ with any n. If $N \in \mathfrak{M}$ is a positive definite kernel, then so is $M \in \mathfrak{M}$ with $M \preceq N$. This is an immediate consequence of the famous Schur theorem on the Schur product of two positive semi-definite matrices. The next proposition says that the geometric mean G is the largest in the order $\preceq$ among means in $\mathfrak{M}$ that are positive definite kernels. When H is a matrix with eigenvalues $s_1, \ldots, s_n \ge 0$, $M(H, H)$ is essentially equal to the Schur multiplication by $[M(s_i, s_j)]_{i,j=1,\cdots,n}$ (up to unitary conjugation, see (3.12)). So one may consider the property (i) below as a generalization of the Schur theorem.

Proposition 3.10. *The following conditions are equivalent for $M \in \mathfrak{M}$:*

(i) M is a Schur multiplier and $M(H, H)X$ is positive if so are $H, X \in B(\mathcal{H})$;
(ii) M is a positive definite kernel;
(iii) $M \preceq G$.

If this is the case, then $\|M(H, K)\|_{(1,1)} \leq \sqrt{\|H\| \times \|K\|}$ for all $H, K \geq 0$.

Proof. (i) $\Rightarrow$ (ii). Choose an orthonormal basis $\{\xi_i\}$. For each n, by setting $X = \sum_{i,j=1}^{n} \xi_i \otimes \xi_j^c$ and $H = \sum_{i=1}^{n} s_i \xi_i \otimes \xi_i^c$ with $s_1, \ldots, s_n \geq 0$, we get

$$M(H, H)X = \sum_{i,j=1}^{n} M(s_i, s_j) \xi_i \otimes \xi_j^c.$$

Hence (i) implies the positive definiteness of $[M(s_i, s_j)]_{i,j=1,\cdots,n}$.
  (ii) $\Rightarrow$ (iii). This is immediate because of

$$\left[\frac{M(s_i, s_j)}{G(s_i, s_j)}\right] = \operatorname{diag}(s_1^{-1/2}, \ldots, s_n^{-1/2})[M(s_i, s_j)]\operatorname{diag}(s_1^{-1/2}, \ldots, s_n^{-1/2})$$

for any $s_1, \ldots, s_n > 0$.
  (iii) $\Rightarrow$ (i). Assume (iii) with the representing measure ν for the ratio $M(e^x, 1)/G(e^x, 1)$. Then Theorem 3.4 implies that M is a Schur multiplier and

$$M(H, H)X = \int_{-\infty}^{\infty} (Hs_H)^{ix} (H^{1/2}XH^{1/2})(Hs_H)^{-ix} d\nu(x),$$

(because of $M(1, 0) = 0$), which is positive if so is X. Furthermore, by Corollary 3.5 we get

$$\||M(H, K)X\|| \leq \||H^{1/2}XK^{1/2}\|| \leq \sqrt{\|H\| \times \|K\|} \, \||X\||$$

for any unitarily invariant norm. Therefore, $\|M(H, K)\|_{(1,1)} \leq \sqrt{\|H\| \times \|K\|}$.
$\square$

3.5 Norm estimates for means

When M is one of A, L and G, it is straight-forward to see $\|M(H, K)\|_{(1,1)} \leq M(\|H\|, \|K\|)$. In fact, this was noticed for G in the proof of Proposition 3.10, and for L we have

$$\||L(H, K)X\|| \leq \int_0^1 \||H^x X K^{1-x}\|| \, dx$$

$$\leq \int_0^1 \|H\|^x \|K\|^{1-x} \, dx \times \||X\|| = L(\|H\|, \|K\|) \, \||X\||$$

for any unitarily invariant norm. As long as $M \preceq M_\infty$ we also get the estimate

$$|||M(H,K)X||| \leq |||M_\infty(H,K)X||| \leq \frac{3}{2}|||HX + XK|||$$

$$\leq \frac{3}{2}(\|H\| + \|K\|) |||X||| \tag{3.16}$$

which is a consequence of Corollary 3.5 and (3.11).

The problem to compute the best possible bound of $\|M(H,K)\|_{(1,1)}$ (in terms of $\|H\|$ and $\|K\|$) is not easy in general. In this section the optimal bound will computed for the mean $M = M_\infty$.

Lemma 3.11. *For every* $s_1, \ldots, s_n \geq 0$,

$$\left\|S_{[s_i \vee s_j]_{i,j=1,\cdots,n}}\right\|_{(1,1)} \leq \frac{2}{\sqrt{3}}\left(\max_i s_i - \min_i s_i\right) + \min_i s_i \leq \frac{2}{\sqrt{3}}\max_i s_i,$$

where $s_i \vee t_j = \max\{s_i, t_j\}$. *Moreover,* $2/\sqrt{3}$ *is the optimal bound in the above estimate.*

Proof. The explicit formula of $\|S_A\|_{(\infty,\infty)}$ for a real 2×2 matrix A was obtained in [21] by using Haagerup's criterion (3.14) and it indeed says

$$\left\|S_{\begin{bmatrix} 1 & 1 \\ 1 & 0 \end{bmatrix}}\right\|_{(\infty,\infty)} = \frac{2}{\sqrt{3}}. \tag{3.17}$$

(In fact, a direct computation of (3.17) with Haagerup's criterion is also easy.) Next, let $s_1, \ldots, s_n \geq 0$. For a permutation γ on $\{1, 2, \ldots, n\}$ with the corresponding permutation matrix Γ we obviously have

$$S_{[s_{\gamma(i)} \vee s_{\gamma(j)}]}(X) = \Gamma\big(S_{[s_i \vee s_j]}(\Gamma^{-1}X\Gamma)\big)\Gamma^{-1}.$$

Thus, we may and do assume $s_1 \geq s_2 \geq \cdots \geq s_n \geq 0$, and the matrix $[s_i \vee s_j]$ can be written as

$$[s_i \vee s_j] = (s_1 - s_2)J_1^{(n)} + (s_2 - s_3)J_2^{(n)} + \cdots + (s_{n-1} - s_n)J_{n-1}^{(n)} + s_n J_n^{(n)},$$

where

$$J_k^{(n)} = \begin{bmatrix} 1 & \cdots & 1 & 1 & \cdots & 1 \\ \vdots & \ddots & \vdots & \vdots & \ddots & \vdots \\ 1 & \cdots & 1 & 1 & \cdots & 1 \\ 1 & \cdots & 1 & 0 & \cdots & 0 \\ \vdots & \ddots & \vdots & \vdots & \ddots & \vdots \\ 1 & \cdots & 1 & 0 & \cdots & 0 \end{bmatrix} \quad \text{(the zero block is } (n-k) \times (n-k)\text{)}.$$

According to (3.17) and Haagerup's criterion, there are $u_1, u_2, v_1, v_2 \in \mathbf{C}^2$ such that $\|u_i\|^2, \|v_j\|^2 \leq 2/\sqrt{3}$ and

$$(u_1, v_1) = (u_1, v_2) = (u_2, v_1) = 1, \quad (u_2, v_2) = 0.$$

For $k = 1, \ldots, n-1$ we get $J_k^{(n)} = [(\xi_i, \eta_j)]$ when $\xi_1 = \cdots = \xi_k = u_1$, $\xi_{k+1} = \cdots = \xi_n = u_2$, $\eta_1 = \cdots = \eta_k = v_1$ and $\eta_{k+1} = \cdots = \eta_n = v_2$. This implies

$$\|S_{J_k^{(n)}}\|_{(\infty,\infty)} \leq \frac{2}{\sqrt{3}} \quad (\text{for } k = 1, \ldots, n-1),$$

and obviously $\|S_{J_n^{(n)}}\|_{(\infty,\infty)} = 1$. Since

$$S_{[s_i \vee s_j]} = (s_1 - s_2)S_{J_1^{(n)}} + (s_2 - s_3)S_{J_2^{(n)}} + \cdots + (s_{n-1} - s_n)S_{J_{n-1}^{(n)}} + s_n S_{J_n^{(n)}}$$

with positive coefficients, we get

$$\|S_{[s_i \vee s_j]}\|_{(\infty,\infty)} \leq \frac{2}{\sqrt{3}}(s_1 - s_n) + s_n$$

as desired. Finally the optimality of $2/\sqrt{3}$ is clear from (3.17). □

The next theorem is a consequence of Lemmas 3.11 and 3.9 (for $\phi = M_\infty$) together with Corollary 3.5 (or Theorem 3.7).

Theorem 3.12. *If $M \in \mathfrak{M}$ satisfies $M \preceq M_\infty$, then*

$$\|M(H,K)\|_{(1,1)} \leq \frac{2}{\sqrt{3}} \max\{\|H\|, \|K\|\}$$

for all $H, K \geq 0$. Consequently, for any unitarily invariant norm $\|\|\cdot\|\|$ we have

$$\|\|M(H,K)X\|\| \leq \frac{2}{\sqrt{3}} \max\{\|H\|, \|K\|\} \|\|X\|\|$$

for all $X \in B(\mathcal{H})$.

For each mean $M \in \mathfrak{M}$ one can define the mean $M^{(-)} \in \mathfrak{M}$ dual to M by

$$M^{(-)}(s,t) = M(s^{-1}, t^{-1})^{-1} \quad \text{for } s, t > 0 \tag{3.18}$$

(see [39, §1]). For $M, N \in \mathfrak{M}$ note that $M \preceq N$ is equivalent to $N^{(-)} \preceq M^{(-)}$. For example, $G^{(-)} = G$, $A^{(-)} = M_{\text{har}}$ and $M_\infty^{(-)} = M_{-\infty}$ concerning means in (3.4). It is easy to see that if H, K are invertible positive operators, then

$$M^{(-)}(H^{-1}, K^{-1})(M(H,K)X)$$
$$= M(H,K)(M^{(-)}(H^{-1}, K^{-1})X) = X \tag{3.19}$$

for all $X \in C_2(\mathcal{H})$. Indeed, this is the application of function calculus to the equality $M^{(-)}(s^{-1}, t^{-1})M(s,t) = 1$. Whenever both M and $M^{(-)}$ are Schur multipliers, (3.19) remains valid for all $X \in B(\mathcal{H})$ so that $M^{(-)}(H^{-1}, K^{-1})$ is the inverse of $M(H,K)$ on $B(\mathcal{H})$. Hence Theorem 3.12 implies

Proposition 3.13. *If* $M \in \mathfrak{M}$ *satisfies* $M_{-\infty} \preceq M \preceq M_{\infty}$ *and* H, K *are invertible positive operators, then*

$$|||M(H, K)X||| \geq \frac{\sqrt{3}}{2} \min\{\|H^{-1}\|^{-1}, \|K^{-1}\|^{-1}\} |||X|||$$

for all unitarily invariant norms and all $X \in B(\mathcal{H})$.

Remark 3.14. The "mean transform" $M(H, K)$ (when $M \preceq M_{\infty}$ for example) sends $\mathcal{I}_{|||\cdot|||}$ (and $\mathcal{I}^{(0)}_{|||\cdot|||}$) into itself (see Propositions 2.7 and 3.3). However, if H, K are positive compact operators in some Schatten class, then one can do better. For example, let us assume $H, K \in \mathcal{C}_{p_0}(\mathcal{H})$ $(1 \leq p_0 \leq \infty)$ and $M = A$, the arithmetic mean. Then, thanks to the (generalized) Hölder inequality

$$\|XY\|_{p_2} \leq \|X\|_{p_1} \|Y\|_{p_0} \quad \left(\text{with } p_1^{-1} + p_0^{-1} = p_2^{-1}\right), \tag{3.20}$$

$M(H, K)$ sends the Schatten class $\mathcal{C}_{p_1}(\mathcal{H})$ into the smaller one $\mathcal{C}_{p_2}(\mathcal{H})$ with the norm bound

$$\|A(H, K)X\|_{p_2} \leq \frac{1}{2}(\|HX\|_{p_2} + \|XK\|_{p_2})$$

$$\leq \frac{1}{2}(\|H\|_{p_0} + \|K\|_{p_0})\|X\|_{p_1} \leq \max\{\|H\|_{p_0}, \|K\|_{p_0}\}\|X\|_{p_1}.$$

We point out that this is a general phenomenon. Namely, let us assume $M \preceq M_{\infty}$ and $p_1^{-1} + p_0^{-1} = p_2^{-1}$ $(1 \leq p_0, p_1, p_2 \leq \infty)$. If positive operators H, K belong to $\mathcal{C}_{p_0}(\mathcal{H})$, then $M(H, K)$ is a bounded linear operator from $\mathcal{C}_{p_1}(\mathcal{H})$ into $\mathcal{C}_{p_2}(\mathcal{H})$ satisfying

$$\|M(H, K)X\|_{p_2} \leq 3\max\{\|H\|_{p_0}, \|K\|_{p_0}\}\|X\|_{p_1}.$$

In fact, the general estimate (3.16) gives

$$\|M(H, K)X\|_{p_2} \leq \frac{3}{2}\|HX + XK\|_{p_2} \leq \frac{3}{2}(\|HX\|_{p_2} + \|XK\|_{p_2})$$

so that the assertion follows from (3.20) as before.

3.6 Kernel and range of $M(H, K)$

Assume $M_{-\infty} \preceq M \preceq M_{\infty}$. When both of $H, K \geq 0$ are invertible, the mean transform $M(H, K) : B(\mathcal{H}) \to B(\mathcal{H})$ is bijective due to (3.19) (for each $X \in B(\mathcal{H})$). In this section we determine the kernel and the closure of the range for general positive H, K.

Lemma 3.15. *Assume that* $M \in \mathfrak{M}$ *satisfies* $M_{-\infty} \preceq M \preceq M_{\infty}$, *and let* H *be a non-singular positive operator.*

(i) If $X \in B(\mathcal{H})$ and $M(H, H)X = 0$, then $X = 0$.
(ii) The range of $M(H, H)$ is dense in $B(\mathcal{H})$ in the strong operator topology.

Proof. (i) For $\delta > 0$ we note

$$0 = E_{(\delta,\infty)}(M(H, H)X)E_{(\delta,\infty)} = M(HE_{(\delta,\infty)}, HE_{(\delta,\infty)})(E_{(\delta,\infty)}XE_{(\delta,\infty)})$$

with the spectral projection $E_{(\delta,\infty)}$ of H. Here, the second equality easily follows from the integral expression pointed out in Remark 2.5, (ii). By restricting everything to the subspace $E_{(\delta,\infty)}\mathcal{H}$ (where $HE_{(\delta,\infty)}$ is an invertible operator), from Proposition 3.13 (and (3.19)) we get $E_{(\delta,\infty)}XE_{(\delta,\infty)} = 0$. We then see $X = 0$ because the non-singularity of H yields the strong convergence $E_{(\delta,\infty)} \nearrow 1$ (as $\delta \searrow 0$).

(ii) Choose and fix $X \in B(\mathcal{H})$ and $\delta > 0$ at first. As above we regard $E_{(\delta,\infty)}XE_{(\delta,\infty)}$ and $HE_{(\delta,\infty)}$ ($\geq \delta$) as operators on $E_{(\delta,\infty)}\mathcal{H}$. Then, the operator equation

$$M(HE_{(\delta,\infty)}, HE_{(\delta,\infty)})Y = E_{(\delta,\infty)}XE_{(\delta,\infty)}$$

for an unknown operator $Y \in B(E_{(\delta,\infty)}\mathcal{H})$ possesses a solution, i.e.,

$$Y = M^{(-)}((HE_{(\delta,\infty)})^{-1}, (HE_{(\delta,\infty)})^{-1})(E_{(\delta,\infty)}XE_{(\delta,\infty)}) \quad \text{(see (3.19))}.$$

However, since $Y \in B(E_{(\delta,\infty)}\mathcal{H})$ $(\subseteq B(\mathcal{H}))$, we observe

$$M(HE_{(\delta,\infty)}, HE_{(\delta,\infty)})Y = M(H, H)Y$$

once again based on the expression in Remark 2.5, (ii). Consequently we have $M(H, H)Y = E_{(\delta,\infty)}XE_{(\delta,\infty)}$, meaning that $E_{(\delta,\infty)}XE_{(\delta,\infty)}$ sits in the range of $M(H, H)$. We thus get the conclusion by letting $\delta \searrow 0$. □

Theorem 3.16. *Assume that $M \in \mathfrak{M}$ satisfies $M_{-\infty} \preceq M \preceq M_\infty$, and let H, K be positive operators.*

I. *Case $M(1, 0) = 0$.*
   (i) *For $X \in B(\mathcal{H})$ we have $M(H, K)X = 0$ if and only if $s_H X s_K = 0$.*
   (ii) *The closure of the range of $M(H, K)$ in the strong operator topology is $s_H B(\mathcal{H}) s_K$.*
II. *Case $M(1, 0) > 0$.*
   (iii) *For $X \in B(\mathcal{H})$ we have $M(H, K)X = 0$ if and only if*

$$s_H X s_K = s_H X(1 - s_K) = (1 - s_H)X s_K = 0.$$

   (iv) *The closure of the range of $M(H, K)$ in the strong operator topology is*

$$\{X \in B(\mathcal{H}) : (1 - s_H)X(1 - s_K) = 0\}.$$

Proof. We begin with the special case $H = K$. We recall

$$M(H, H)X$$
$$= s_H(M(H, H)X)s_H + M(1, 0)(HX(1 - s_H) + (1 - s_H)XH)$$
$$= M(Hs_H, Hs_H)(s_H X s_H) + M(1, 0)(HX(1 - s_H) + (1 - s_H)XH)$$

(see Lemma 2.9 and (3.10)). By restricting everything to the subspace $s_H \mathcal{H}$ (where Hs_H is non-singular) Lemma 3.15, (i) says $M(Hs_H, Hs_H)(s_H X s_H) = 0$ if and only if $s_H X s_H = 0$, showing (i). When $M(1, 0) > 0$, the additional requirement $HX(1 - s_H) = (1 - s_H)XH = 0$ is needed. However, this is obviously equivalent to $s_H X(1 - s_H) = (1 - s_H)X s_H = 0$, which corresponds to (iii). On the other hand, from Lemma 3.15, (ii) (and the above decomposition) we easily get (ii) and (iv). Note that to show (iv) we need the following obvious fact for instance: $HB(\mathcal{H})(1 - s_H)$ is strongly dense in

$$\{X \in B(\mathcal{H}) : s_H X s_H = (1 - s_H)X s_H = (1 - s_H)X(1 - s_H) = 0\},$$

i.e., operators with only (non-zero) "(1, 2)-components".

In the rest of the proof we will deal with the general case. With $\tilde{H}, \tilde{X}$ in Remark 2.12 we have

$$M(H, K)X = 0 \Longleftrightarrow M(\tilde{H}, \tilde{H})\tilde{X} = 0,$$

which is also equivalent to $s_{\tilde{H}} \tilde{X} s_{\tilde{H}} = 0$ (with the additional requirement $s_{\tilde{H}} \tilde{X}(1 - s_{\tilde{H}}) = (1 - s_{\tilde{H}})\tilde{X} s_{\tilde{H}} = 0$ when $M(1, 0) > 0$) from the first part of the proof. But, since $s_{\tilde{H}} = \begin{bmatrix} s_H & 0 \\ 0 & s_K \end{bmatrix}$, we easily get (i) and (iii) (in the general setting). Indeed, we have

$$s_{\tilde{H}} \tilde{X} s_{\tilde{H}} = 0 \Longleftrightarrow s_H X s_K = 0,$$
$$s_{\tilde{H}} \tilde{X}(1 - s_{\tilde{H}}) = 0 \Longleftrightarrow s_H X(1 - s_K) = 0,$$
$$(1 - s_{\tilde{H}})\tilde{X}(1 - s_{\tilde{H}}) = 0 \Longleftrightarrow (1 - s_H)X(1 - s_K) = 0.$$

To investigate the range, we consider the projections

$$P_1 = \begin{bmatrix} 1 & 0 \\ 0 & 0 \end{bmatrix}, \quad P_2 = \begin{bmatrix} 0 & 0 \\ 0 & 1 \end{bmatrix} \quad (\text{in } B(\mathcal{H} \oplus \mathcal{H})).$$

The range $M(H, K)(B(\mathcal{H}))$ is $P_1(M(\tilde{H}, \tilde{H})(B(\mathcal{H} \oplus \mathcal{H}))P_2$ (see Remark 2.12) with the natural identification of the $(1, 2)$-corner of $B(\mathcal{H} \oplus \mathcal{H})$ with $B(\mathcal{H})$. We claim

$$\overline{P_1(M(\tilde{H}, \tilde{H})(B(\mathcal{H} \oplus \mathcal{H}))P_2} = P_1\overline{(M(\tilde{H}, \tilde{H})(B(\mathcal{H} \oplus \mathcal{H}))}P_2.$$

At first, $\supseteq$ is obvious. To see $\subseteq$, we choose and fix Y from the left-hand side. We note $Y = P_1 Y P_2$ and can choose $Y_\lambda = M(\tilde{H}, \tilde{H})Z_\lambda$ (for some $Z_\lambda \in B(\mathcal{H} \oplus \mathcal{H})$) such that $P_1 Y_\lambda P_2 \to Y$ strongly. But notice

$$P_1 Y_\lambda P_2 = P_1(M(\tilde{H}, \tilde{H}) Z_\lambda) P_2 = M(\tilde{H}, \tilde{H})(P_1 Z_\lambda P_2)$$

due to the fact that P_1 and P_2 commute with $\tilde{H}$ (recall the integral expression in Remark 2.12). Therefore, each $P_1 Y_\lambda P_2$ actually belongs to the range $\overline{M(\tilde{H}, \tilde{H})(B(\mathcal{H} \oplus \mathcal{H}))}$ so that the limit Y sits in the strong closure $\overline{M(\tilde{H}, \tilde{H})(B(\mathcal{H} \oplus \mathcal{H}))}$. Hence, we have

$$Y = P_1 Y P_2 \in P_1 \overline{M(\tilde{H}, \tilde{H})(B(\mathcal{H} \oplus \mathcal{H}))} P_2,$$

and the claim is established.

From the discussions so far we have

$$\overline{M(H, K)(B(\mathcal{H}))} = \overline{P_1(M(\tilde{H}, \tilde{H})(B(\mathcal{H} \oplus \mathcal{H}))} P_2$$
$$= P_1 \overline{(M(\tilde{H}, \tilde{H})(B(\mathcal{H} \oplus \mathcal{H}))} P_2. \qquad (3.21)$$

When $M(1, 0) = 0$, we have

$$\overline{M(H, K)(B(\mathcal{H}))} = P_1 s_{\tilde{H}} B(\mathcal{H} \oplus \mathcal{H}) s_{\tilde{H}} P_2$$
$$= s_{\tilde{H}} P_1 B(\mathcal{H} \oplus \mathcal{H}) P_2 s_{\tilde{H}} = s_{\tilde{H}} B(\mathcal{H}) s_{\tilde{H}}.$$

Here, the first equality follows from (3.21) and the first part of the proof (i.e, (ii) in the special case $H = K$) while the second is a consequence of the commutativity of P_1, P_2 with $\tilde{H}$. The last equality comes from the above-mentioned natural identification. We note that the $B(\mathcal{H})$ (appearing in the far right side) is the one sitting at the $(1, 2)$-corner so that $s_{\tilde{H}} B(\mathcal{H}) s_{\tilde{H}}$ actually means $s_H B(\mathcal{H}) s_K$ (sitting at the same place). Therefore, we have shown (ii).

On the other hand, when $M(1, 0) > 0$, from (3.21) (and (iv) in the special case) we similarly get

$$\overline{M(H, K)(B(\mathcal{H}))}$$
$$= P_1 \Big(s_{\tilde{H}} B(\mathcal{H} \oplus \mathcal{H}) s_{\tilde{H}}$$
$$\qquad + (1 - s_{\tilde{H}}) B(\mathcal{H} \oplus \mathcal{H}) s_{\tilde{H}} + s_{\tilde{H}} B(\mathcal{H} \oplus \mathcal{H})(1 - s_{\tilde{H}}) \Big) P_2$$
$$= s_{\tilde{H}} P_1 B(\mathcal{H} \oplus \mathcal{H}) P_2 s_{\tilde{H}}$$
$$\qquad + (1 - s_{\tilde{H}}) P_1 B(\mathcal{H} \oplus \mathcal{H}) P_2 s_{\tilde{H}} + s_{\tilde{H}} P_1 B(\mathcal{H} \oplus \mathcal{H}) P_2 (1 - s_{\tilde{H}})$$
$$= s_{\tilde{H}} B(\mathcal{H}) s_{\tilde{H}} + (1 - s_{\tilde{H}}) B(\mathcal{H}) s_{\tilde{H}} + s_{\tilde{H}} B(\mathcal{H})(1 - s_{\tilde{H}}).$$

The $B(\mathcal{H})$ appearing at the end is once again the one at the $(1, 2)$-corner, and the same reasoning as in the last part of the preceding paragraph yields (iv) in the general case. $\square$

3.7 Notes and references

1. Means of operators

In [39] the class $\mathfrak{M}$ (in Definition 3.1) of homogeneous symmetric means was introduced, and for matrices H, K, X (with $H, K \geq 0$) and $M \in \mathfrak{M}$ the matrix mean $M(H, K)X$ was defined by (1.1). With this definition Theorem 3.7 was obtained for matrices (as [39, Theorem 1.1]), and many norm inequalities were obtained. We cannot determine if every $M \in \mathfrak{M}$ is a Schur multiplier (probably not), and this problem seems to deserve further investigation. Anyway the criterion $M \preceq M_\infty$ obtained in Proposition 3.3, (b) is good enough in almost all circumstances. The implication (iv) $\Rightarrow$ (ii) in Theorem 3.7 (at least in the matrix case, or equivalently (3.15)) has been known to many specialists ([42, p. 343] and [4, p. 363] for example) and indeed used as a standard tool for showing norm inequalities. We actually have the bi-implication here. Therefore, the theorem can be also used to check failure of certain norm inequalities, which will be carried out in our forthcoming article [55]. In §8.1 and §A.1 we will deal with "operator means" $M(H, K)X$ for functions M in wider classes. This will make it possible to study norm inequalities for certain operators which are not operator means in the sense of the present chapter.

Our theory of operator means is useful in study of certain operator equations. Let us assume the invertibility of $H, K \geq 0$ for simplicity and regard

$$M(H, K)X = Y$$

as an operator equation with an unknown operator X. Then, (3.18) and (3.19) show that $X = M^{(-)}(H^{-1}, K^{-1})Y$ gives rise to a solution. With this idea concrete integral expressions for solutions to many operator equations were obtained in [39, §4]. In [68] related analysis was also made by G. K. Pedersen from the viewpoint of "operator differentials" (see also [33, 67]). Theorem 3.16 in §3.6 provides us useful information on uniqueness of solutions to the above operator equation.

Another important notion of operator means, quite different from those treated in the present monograph, is the one axiomatically introduced by F. Kubo and T. Ando in [57]. An operator mean in their sense is a binary operation $B(\mathcal{H})_+ \times B(\mathcal{H})_+ \to B(\mathcal{H})_+$, and it bijectively corresponds to an operator monotone function on $\mathbf{R}_+$. For example, the geometric mean (formerly introduced by W. Pusz and L. Woronowicz in [73]) is given as $H \# K = H^{\frac{1}{2}}(H^{-\frac{1}{2}} K H^{-\frac{1}{2}})^{\frac{1}{2}} H^{\frac{1}{2}}$ for positive invertible $H, K \in B(\mathcal{H})$ while our geometric mean $G(H, K)X = H^{\frac{1}{2}} X K^{\frac{1}{2}}$ is no longer positive even when $X = 1$.

2. Arithmetic-geometric mean inequality and related topics

The arithmetic-geometric mean inequality (1.4) for unitarily invariant norms was first noticed by R. Bhatia and C. Davis in [10], and its alternative proofs (and/or some discussions) were worked out by many authors including

R. A. Horn ([41]), F. Kittaneh ([50, 51]), R. Mathias ([63]) and probably some others. Proofs presented in [41, 63] are indeed based on the method explained in **1**. The article [13] by R. Bhatia and K. Parthasarathy is closely related to our previous works [38, 39, 54], and this method was systematically used to derive an abundance of known and new norm inequalities. The same method was used by X. Zhan ([83, Theorem 6 and Corollary 7]) to show the following generalizations of the arithmetic-geometric mean inequality (as well as the Heinz inequality (1.3)):

(i) for $x \in (-2, 2]$ and $\theta \in [1/4, 3/4]$,

$$\frac{2+x}{2} \, |||H^\theta X K^{1-\theta} + H^{1-\theta} X K^\theta||| \leq |||HX + XK + xH^{1/2}XK^{1/2}|||;$$

(ii) for $x \in (-2, 2]$,

$$(2+x)|||H^{1/2}XK^{1/2}||| \leq |||HX + XK + xH^{1/2}XK^{1/2}|||.$$

Similar results (based on the similar method) were also obtained in [78].

The following inequality was obtained by D. Jocić ([45, Theorem 3.1]) as an application of the arithmetic-geometric mean inequality:

$$||| \, |HX + XK|^p \, ||| \leq 2^{p-1}||X||^{p-1}||| \, |H|^{p-1}HX + XK|K|^{p-1}|||$$

for $p \geq 3$ and self-adjoint operators H, K. It generalizes the earlier result

$$|||(H - K)^{2n+1}||| \leq 2^{2n}|||H^{2n+1} - K^{2n+1}|||$$

due to D. Jocić and F. Kittaneh ([46], and also see [7]). In fact, when $p = 2n+1$ odd, by setting $X = 1$ and using $-K$ instead one gets $|H|^{2n}H = H^{2n+1}$ and $(-K)|(-K)|^{2n} = -K^{2n+1}$. This perturbation estimate in particular shows $H - K \in \mathcal{C}_{(2n+1)p}$ as long as $H^{2n+1} - K^{2n+1} \in \mathcal{C}_p$ and $p \in [1, \infty)$, which improves L. S. Koplienko's result in [52].

G. Corach, H. Porta and L. Recht studied the set of invertible self-adjoint operators (and some other sets) as a space equipped with a certain natural Finsler metric (see [60]). In [19] from the differential geometry viewpoint they arrived at the inequality

$$\|X\| \leq \frac{1}{2}\|HXH^{-1} + H^{-1}XH\|$$

for an invertible self-adjoint operator H. This corresponds to the norm-decreasing property of a certain tangential map, and their proof actually uses Schur products. As noticed in [28, 51] for example (change X to HXH and use the standard 2×2-matrix trick in Remark 2.12), their inequality is nothing but the arithmetic-geometric mean inequality (in the operator norm). In [20] they also gave a geometric interpretation of the Segal inequality

$$\|e^{H+K}\| \leq \|e^{H/2}e^K e^{H/2}\| \ \left(\leq \|e^H e^K\|\right)$$

for self-adjoint operators H, K.

3. Arithmetic-logarithmic-geometric mean inequality

The arithmetic-logarithmic-geometric mean inequality (1.8) (as well as some further extensions such as monotonicity of the norms (1.9) in m and n) was proved in [38]. In [9] R. Bhatia pointed out a close connection between the logarithmic-geometric mean inequality and the Golden-Thompson-type norm inequality (extending the Segal inequality)

$$|||e^{H+K}||| \leq |||e^H e^K|||$$

for self-adjoint operators H, K based on the differential geometry viewpoint (akin to [19, 20]). (See [8, 37, 77] for the Golden-Thompson-type inequality.)

4. Schur multipliers in the matrix case

Haagerup's criterion and (3.14) were presented in his unpublished notes [31, 32], and a proof is available in the literature. Namely, the formula was shown in the article [5] by T. Ando and K. Okubo as a consequence of its variant for the numerical radius norm. The Ando-Okubo theorem was recently extended to $B(\mathcal{H})$ by T. Itoh and M. Nagisa in [44].

Materials in §3.3 are somewhat technical. But, we need them (especially Lemma 3.9) to reduce the proof of Theorem 3.12 in §3.5 to the matrix case. In fact, this technique enables us to make use of Lemma 3.11 (based on (3.14)).

(Sub)majorization theory for eigenvalues and singular values of matrices provides a powerful tool in study of matrix (also operator) norm inequalities for unitarily invariant norms (see [34, 62] and also [1, 2, 8] for surveys on recent results). Among others, T. Ando, R. A. Horn and C. R. Johnson obtained in [4] a fundamental majorization for singular values of Hadamard (or Schur) products of matrices, which implies (3.15) as a corollary. Majorization method was implicitly used in the proof of Proposition 2.6; however it does not have much to do with the present monograph.

4

Convergence of means

In this chapter we will investigate continuity properties of means (in operator variables). In fact, the convergence $M(H_n, K_n)X \to M(H, K)X$ in a unitarily invariant norm is discussed under the strong convergence $H_n \to H$, $K_n \to K$. Our main result here is Theorem 4.1 in §4.1, and some related convergence results are also presented in §4.2 as variants of (the proof of) the main theorem.

4.1 Main convergence result

Norm convergence is guaranteed under many circumstances. Although the conditions imposed in the theorem below may not be optimal, many practical situations are being covered.

Theorem 4.1. *Let $M \in \mathfrak{M}$ be such that $M \preceq M_\infty$, and $||| \cdot |||$ be a unitarily invariant norm. Let H, K, H_n and K_n ($n = 1, 2, \dots$) be positive operators such that $H_n \to H$ and $K_n \to K$ in the strong operator topology. Assume in addition one of the following assumptions:*

(a) $||| \cdot |||$ *is dominated by* $\| \cdot \|_2$,
(b) $s_{H_n} \to s_H$ *and* $s_{K_n} \to s_K$ *strongly,*
(c) $M \preceq L$, *where L denotes the logarithmic mean.*

Then we have

$$\lim_{n \to \infty} |||M(H_n, K_n)X - M(H, K)X||| = 0$$

for all $X \in \mathcal{I}_{|||\cdot|||}^{(0)}$.

Proof. Thanks to the assumption $M \preceq M_\infty$ and the boundedness of $\|H_n\|$ and $\|K_n\|$, Theorem 3.12 implies that there is a $\kappa < \infty$ such that

$$|||M(H_n, K_n)X||| \leq \kappa |||X|||$$

for all $n = 1, 2, \ldots$ and all $X \in B(\mathcal{H})$. Since $\mathcal{I}_{\text{fin}}$ is dense in $\mathcal{I}^{(0)}_{|||\cdot|||}$, it suffices to show the required norm convergence for rank-one operators X.

Case (a). This case is immediately seen because $M(H_n, K_n) \to M(H, K)$ strongly as operators acting on the Hilbert-Schmidt class $\mathcal{C}_2(\mathcal{H})$ (see the proof (iv) $\Rightarrow$ (iii) of Theorem 3.7).

Case (b). Any unitarily invariant norm is dominated by $\|\cdot\|_1$, and hence we may prove the case $|||\cdot||| = \|\cdot\|_1$. Put $\tilde{H}_n = H_n + (1 - s_{H_n})$ and $\tilde{H} = H + (1 - s_H)$, so $\tilde{H}_n$ and $\tilde{H}$ are non-singular positive operators. Since $\tilde{H}_n \to \tilde{H}$ strongly, it is well-known that $\tilde{H}_n^{ix} \to \tilde{H}^{ix}$ strongly for all $x \in \mathbf{R}$. Hence

$$(H_n s_{H_n})^{ix} = \tilde{H}_n^{ix} s_{H_n} \longrightarrow \tilde{H}^{ix} s_H = (H s_H)^{ix}$$

strongly for all $x \in \mathbf{R}$. Similarly, $(K_n s_{K_n})^{ix} \to (K s_K)^{ix}$ strongly. For a rank-one operator X, we claim that

$$\lim_{n \to \infty} \| M_\infty(H_n, K_n)X - M_\infty(H, K)X \|_1 = 0. \tag{4.1}$$

In fact, (3.11) shows

$$M_\infty(H_n, K_n)X = -\frac{1}{2} \int_{-\infty}^{\infty} (H_n s_{H_n})^{ix}(H_n X + X K_n)(K_n s_{K_n})^{-ix} f(x)\, dx$$
$$+ H_n X + X K_n,$$
$$M_\infty(H, K)X = -\frac{1}{2} \int_{-\infty}^{\infty} (H s_H)^{ix}(HX + XK)(K s_K)^{-ix} f(x)\, dx$$
$$+ HX + XK.$$

It is straight-forward to see $\|(H_n X + X K_n) - (HX + XK)\|_1 \to 0$ since X is of rank-one. So it suffices to show

$$\lim_{n \to \infty} \| (H_n s_{H_n})^{ix}(H_n X + X K_n)(K_n s_{K_n})^{-ix}$$
$$- (H s_H)^{ix}(HX + XK)(K s_K)^{-ix} \|_1 = 0 \tag{4.2}$$

for all $x \in \mathbf{R}$. Indeed, we can then apply Theorem A.5 and the Lebesgue dominated convergence theorem to get (4.1). However, the $\|\cdot\|_1$-norm in (4.2) is majorized by

$$\|(H_n s_{H_n})^{ix}((H_n X + X K_n) - (HX + XK))(K_n s_{K_n})^{-ix}\|_1$$
$$+ \|((H_n s_{H_n})^{ix} - (H s_H)^{ix})(XH + XK)(K_n s_{K_n})^{-ix}\|_1$$
$$+ \|(H s_H)^{ix}(HX + XK)((K_n s_{K_n})^{-ix} - (K s_K)^{-ix})\|_1$$
$$\leq \|(H_n X + X K_n) - (HX + XK)\|_1$$
$$+ \|((H_n s_{H_n})^{ix} - (H s_H)^{ix})(XH + XK)\|_1$$
$$+ \|(HX + XK)((K_n s_{K_n})^{-ix} - (K s_K)^{-ix})\|_1$$

so that (4.2) is obtained from the strong convergence

$$(H_n s_{H_n})^{ix} \longrightarrow (H s_H)^{ix}, \quad (K_n s_{K_n})^{-ix} \longrightarrow (K s_K)^{-ix}.$$

When $M \preceq M_\infty$ Theorem 3.4 (see (3.8)) guarantees

$$M(H_n, K_n)X = \int_{-\infty}^{\infty} (H_n s_{H_n})^{ix} (M_\infty(H_n, K_n)X)(K_n s_{K_n})^{-ix} d\nu(x)$$
$$+ M(1,0)(s_{H_n} X(1 - s_{K_n}) + (1 - s_{H_n})X s_{K_n}),$$
$$M(H, K)X = \int_{-\infty}^{\infty} (H s_H)^{ix} (M_\infty(H, K)X)(K s_K)^{-ix} d\nu(x)$$
$$+ M(1,0)(s_H X(1 - s_K) + (1 - s_H)X s_K).$$

The strong convergence $s_{H_n} \to s_H$, $s_{K_n} \to s_K$ is assumed while the preceding claim says (4.1). Therefore, by making use of these we can repeat the arguments in the proof of the claim for the above $M(H_n, K_n)X$ and $M(H, K)X$ to conclude

$$\|M(H_n, K_n)X - M(H, K)X\|_1 = 0.$$

Case (c). As usual we may and do assume $H_n = K_n$ and $H = K$ thanks to the 2×2-matrix trick, and we set $\alpha = \sup_n \|H_n\|$ ($< \infty$). Choose and fix $\delta > 0$. Let us assume $E_{\{\delta\}}(H) = 0$ (where $E_\Lambda(H)$ denotes the spectral measure for H) so that we have the strong convergence

$$P_n = E_{[0,\delta)}(H_n) \longrightarrow P = E_{[0,\delta)}(H)$$

(see [74, Theorem VIII.24]). We consider the decomposition

$$H_n = H_n P_n + H_n P_n^\perp, \quad H = HP + HP^\perp.$$

Based on the integral expression in Remark 2.5, (ii) we easily have

$$M(H_n, H_n)X = M(H_n P_n, H_n P_n)(P_n X P_n) + M(H_n P_n, H_n P_n^\perp)(P_n X P_n^\perp)$$
$$+ M(H_n P_n^\perp, H_n P_n)(P_n^\perp X P_n) + M(H_n P_n^\perp, H_n P_n^\perp)(P_n^\perp X P_n^\perp)$$

(and the similar decomposition of $M(H, H)X$).

We recall the general fact

$$\||L(H, K)X\|| \leq L(\|H\|, \|K\|) \, \||X\||$$

(see the paragraph before Lemma 3.11). Corollary 3.5 together with this implies

$$\||M(H_n P_n, H_n P_n^\perp)(P_n X P_n^\perp)\|| \leq \||L(H_n P_n, H_n P_n^\perp)(P_n X P_n^\perp)\||$$
$$\leq L(\|H_n P_n\|, \|H_n P_n^\perp\|) \, \||P_n X P_n^\perp\||$$
$$\leq L(\delta, \alpha) \, \||X\|| \tag{4.3}$$

thanks to $\|H_n P_n\| \leq \delta$, $\|H_n P_n^\perp\| \leq \alpha$. Of course the same estimate is available for

$$|||M(H_n P_n^\perp, H_n P_n)(P_n^\perp X P_n)|||, \quad |||M(HP, HP^\perp)(PXP^\perp)|||$$
$$\text{and} \quad |||M(HP^\perp, HP)(P^\perp XP)|||.$$

Similarly we have

$$\begin{cases} |||M(H_n P_n, H_n P_n)(P_n X P_n)||| \le \delta |||X|||, \\ |||M(HP, HP)(PXP)||| \le \delta |||X|||. \end{cases} \qquad (4.4)$$

The estimates so far imply

$$|||M(H_n, H_n)X - M(H, H)X|||$$
$$\le |||M(H_n P_n^\perp, H_n P_n^\perp)(P_n^\perp X P_n^\perp) - M(HP^\perp, HP^\perp)(P^\perp X P^\perp)|||$$
$$+ (2\delta + 4L(\delta, \alpha))|||X|||. \qquad (4.5)$$

Note $L(\delta, \alpha) \searrow 0$ as $\delta \searrow 0$. For each $\varepsilon > 0$, we can choose $\delta > 0$ such that

$$(2\delta + 4L(\delta, \alpha))|||X||| \le \varepsilon \quad \text{and} \quad E_{\{\delta\}}(H) = 0$$

(due to the separability of our Hilbert space). Then, we have

$$|||M(H_n, H_n)X - M(H, H)X|||$$
$$\le |||M(H_n P_n^\perp, H_n P_n^\perp)(P_n^\perp X P_n^\perp) - M(HP^\perp, HP^\perp)(P^\perp X P^\perp)||| + \varepsilon.$$

Since $E_{\{\delta\}}(H) = 0$, we have the strong convergence

$$s_{H_n P_n^\perp} = P_n^\perp \longrightarrow s_{HP^\perp} = P^\perp, \quad H_n P_n^\perp \longrightarrow HP^\perp$$

as was remarked at the beginning, and Case (b) (or more precisely Remark 4.2, (1) below together with the obvious fact $\lim_{n\to\infty} |||P_n^\perp X P_n^\perp - P^\perp X P^\perp||| = 0$) guarantees

$$\lim_{n\to\infty} |||M(H_n P_n^\perp, H_n P_n^\perp)(P_n^\perp X P_n^\perp) - M(HP^\perp, HP^\perp)(P^\perp X P^\perp)||| = 0.$$

Therefore, we have

$$\limsup_{n\to\infty} |||M(H_n, H_n)X - M(H, H)X||| \le \varepsilon,$$

and the proof is completed. $\square$

Remark 4.2. Some remarks are in order.

(1) The conclusion of Theorem 4.1 can be a bit strengthened: if $X_n, X \in \mathcal{I}_{|||\cdot|||}^{(0)}$ and $|||X_n - X||| \to 0$, then

$$\lim_{n\to\infty} |||M(H_n, K_n)X_n - M(H, K)X||| = 0$$

under the same situation. The result indeed follows from

$$|||M(H_n, K_n)X_n - M(H, K)X|||$$
$$\le |||M(H_n, K_n)(X_n - X)||| + |||M(H_n, K_n)X - M(H, K)X|||$$
$$\le \kappa |||X_n - X||| + |||M(H_n, K_n)X - M(H, K)X|||.$$

(2) The case (a) covers the Schatten p-norm $\|\cdot\|_p$ for $2 \le p < \infty$ and the operator norm $\|\cdot\|$, so if $M \preceq M_\infty$ and $H_n \to H$, $K_n \to K$ strongly, then we have

$$\lim_{n\to\infty} \|M(H_n, K_n)X - M(H, K)X\| = 0$$

for all $X \in \mathcal{C}(\mathcal{H})$ $\left(= \mathcal{I}_{\|\cdot\|}^{(0)}\right)$, the algebra of all compact operators.

(3) The condition (b) is automatic as long as $s_H \ge s_{H_n}$ (for n large enough). Hence, for example when either $H_n \nearrow H$, $K_n \nearrow K$ or H, K are non-singular, the condition (b) is satisfied. In fact, thanks to $s_{H_n} \ge H_n(\varepsilon + H_n)^{-1}$ ($\varepsilon > 0$) and the strong convergence $H_n \to H$ we have

$$(s_H \xi, \xi) \ge \limsup_{n\to\infty}(s_{H_n}\xi, \xi) \ge \liminf_{n\to\infty}(s_{H_n}\xi, \xi)$$

$$\ge \liminf_{n\to\infty}(H_n(\varepsilon + H_n)^{-1}\xi, \xi) = (H(\varepsilon + H)^{-1}\xi, \xi)$$

for each vector ξ. By letting $\varepsilon \searrow 0$ one gets $\lim_{n\to\infty}(s_{H_n}\xi, \xi) = (s_H\xi, \xi)$, showing $s_{H_n} \to s_H$ strongly.

(4) When $M \in \mathfrak{M}$ is a Schur multiplier, one can observe from the argument before Remark 2.5 that $M(H, K)$ on $\mathcal{I}_{\|\cdot\|}$ is the transpose of $M(K, H)$ on $\mathcal{I}_{\|\cdot\|}^{(0)}$ under the duality $\mathcal{I}_{\|\cdot\|} = \left(\mathcal{I}_{\|\cdot\|'}^{(0)}\right)^*$. Here, $\|\cdot\|'$ is the conjugate norm of $\|\cdot\|$, and the duality is given by the bilinear form $(X, Y) \in \mathcal{I}_{\|\cdot\|} \times \mathcal{I}_{\|\cdot\|'}^{(0)} \mapsto \mathrm{Tr}(XY) \in \mathbf{C}$. Hence $M(H, K)$ on $\mathcal{I}_{\|\cdot\|}$ is w*-w*-continuous, that is $\sigma\left(\mathcal{I}_{\|\cdot\|}, \mathcal{I}_{\|\cdot\|'}^{(0)}\right)$-$\sigma\left(\mathcal{I}_{\|\cdot\|}, \mathcal{I}_{\|\cdot\|'}^{(0)}\right)$-continuous, as in Remark 2.5, (i). It is seen from this fact that $M(H_n, K_n)X \to M(H, K)X$ in $\sigma\left(\mathcal{I}_{\|\cdot\|}, \mathcal{I}_{\|\cdot\|'}^{(0)}\right)$ for all $X \in \mathcal{I}_{\|\cdot\|}$ in the situation of (b) or (c) in Theorem 4.1.

4.2 Related convergence results

Variants of the arguments presented in the proof of Theorem 4.1 enable us to obtain some related convergence criteria in many settings. We begin with the strong convergence $M(H_n, K_n)X \to M(H, K)X$, which is somewhat easier to handle.

Proposition 4.3. *Assume that $M \in \mathfrak{M}$ satisfies $M \preceq M_\infty$. Let H, K, H_n and K_n $(n = 1, 2, \dots)$ be positive operators such that $H_n \to H$, $K_n \to K$, $s_{H_n} \to s_H$ and $s_{K_n} \to s_K$ in the strong operator topology. Then, for each $X \in B(\mathcal{H})$, means $M(H_n, K_n)X$ tend to $M(H, K)X$ in the strong operator topology.*

Proof. As remarked in the proof of Theorem 4.1, we have the strong convergence $(H_n s_{H_n})^{ix} \to (H s_H)^{ix}$, $(K_n s_{K_n})^{ix} \to (K s_K)^{ix}$ (for each $x \in \mathbf{R}$). We consider the special case $M = M_\infty$ at first. By substituting the integral expression (3.11) to the right-hand side of the obvious equation

$$\| (M_\infty(H_n, K_n)X - M_\infty(H, K)X)\xi \|$$
$$= \sup_{\|\eta\| \leq 1} | ((M_\infty(H_n, K_n)X - M_\infty(H, K)X)\xi, \eta) |,$$

we easily observe

$$\| (M_\infty(H_n, K_n)X - M_\infty(H, K)X)\xi \|$$
$$\leq \|((H_n X + X K_n) - (HX + XK))\xi \|$$
$$+ \frac{1}{2} \int_{-\infty}^{\infty} \|((H_n s_{H_n})^{ix}(H_n X + X K_n)(K_n s_{K_n})^{-ix}$$
$$- (H s_H)^{ix}(HX + XK)(K s_K)^{-ix})\xi \| f(x)\, dx.$$

Since

$$(H_n s_{H_n})^{ix}(H_n X + X K_n)(K_n s_{K_n})^{-ix} \longrightarrow (H s_H)^{ix}(HX + XK)(K s_K)^{-ix}$$

strongly, the above estimate (together with the Lebesgue dominated convergence theorem) implies the strong convergence

$$M_\infty(H_n, K_n)X \longrightarrow M_\infty(H, K)X.$$

Moreover, since Theorem 3.12 implies the uniform boundedness

$$\sup_n \|M_\infty(H_n, K_n)X\| < \infty, \tag{4.6}$$

the following strong convergence is also valid:

$$(H_n s_{H_n})^{ix}(M_\infty(H_n, K_n)X)(K_n s_{K_n})^{-ix}$$
$$\longrightarrow (H s_H)^{ix}(M_\infty(H, K)X)(K s_K)^{-ix}. \tag{4.7}$$

We now assume $M \preceq M_\infty$. Then, based on Theorem 3.4 (i.e., (3.8)) we obtain the similar estimate for $\|(M(H_n, K_n)X - M(H, K)X)\xi\|$ as above with the integrand

$$\|((H_n s_{H_n})^{ix}(M_\infty(H_n, K_n)X)(K_n s_{K_n})^{-ix}$$
$$- (H s_H)^{ix}(M_\infty(H, K)X)(K s_K)^{-ix})\xi \|.$$

Therefore, (4.6), (4.7) and another use of the Lebesgue dominated convergence theorem yield the strong convergence $M(H_n, K_n)X \to M(H, K)X$. □

The strong convergence of $M(H_n, K_n)X$ to $M(H, K)X$ is also guaranteed by (i) the strong convergence $H_n \to H, K_n \to K$, (ii) $X \in C(\mathcal{H})$ (i.e., X is compact) and (iii) $M \preceq L$ (i.e., the condition (c) in Theorem 4.1). We will just sketch the arguments, and full details are left to the reader. In fact, by using the same decomposition (as well as the notations) as in the proof of Theorem 4.1, (c) and the estimates (4.3), (4.4) for $||| \cdot ||| = \| \cdot \|$ the operator norm, we obtain the following estimate for each vector ξ:

$$\| (M(H_n, H_n)X - M(H, H)X)\,\xi\|$$
$$\leq \left\|\left(M(H_n P_n^{\perp}, H_n P_n^{\perp})(P_n^{\perp} X P_n^{\perp}) - M(HP^{\perp}, HP^{\perp})(P^{\perp} X P^{\perp})\right)\xi\right\|$$
$$+(2\delta + 4L(\delta, \alpha))\|X\| \times \|\xi\|.$$

Therefore, we can repeat the arguments at the end of the part (c) in the proof of Theorem 4.1 to get the desired convergence; in fact, use the above estimate in place of (4.5) and apply Proposition 4.3 (see also Remark 4.2, (1)) together with $\|P_n^{\perp} X P_n^{\perp} - P^{\perp} X P^{\perp}\| \to 0$, which is a consequence of the compactness of X.

We point out that the arguments in Case (b) in Theorem 4.1 gives us the norm convergence $M(H_n, K_n)X \to M(H, K)X$ valid for all $X \in \mathcal{I}_{|||\cdot|||}$ (instead of $\mathcal{I}_{|||\cdot|||}^{(0)}$ under a stronger condition).

Proposition 4.4. *Assume that $M \in \mathfrak{M}$ satisfies $M \preceq M_{\infty}$. Let H, K, H_n and K_n $(n = 1, 2, \ldots)$ be positive operators such that H, K are invertible, $\|H_n - H\| \to 0$ and $\|K_n - K\| \to 0$. Then for any unitarily invariant norm $|||\cdot|||$ we have*

$$\lim_{n \to \infty} |||M(H_n, K_n)X - M(H, K)X||| = 0$$

for all $X \in \mathcal{I}_{|||\cdot|||}$. In particular,

$$\lim_{n \to \infty} \|M(H_n, K_n)X - M(H, K)X\| = 0$$

for all $X \in B(\mathcal{H})$.

Proof. Note that H_n, K_n are invertible for large n and $\|H_n^{ix} - H^{ix}\| \to 0$, $\|K_n^{ix} - K^{ix}\| \to 0$ for all $x \in \mathbf{R}$. By using the expression (3.11) (together with Theorem A.5 and the Lebesgue dominated convergence theorem) it is easy to see that

$$\lim_{n \to \infty} |||M_{\infty}(H_n, K_n)X - M_{\infty}(H, K)X||| = 0$$

for all $X \in \mathcal{I}_{|||\cdot|||}$. Next, by using the expression

$$M(H_n, K_n)X = \int_{-\infty}^{\infty} H_n^{ix}(M_{\infty}(H_n, K_n)X)K_n^{-ix}d\nu(x)$$

and the same for H, K we obtain the conclusion. $\square$

A-L-G interpolation means M_α

Three special one-parameter families of symmetric homogeneous means were investigated in our previous article [39]: *A-L-G* interpolation means M_α, Heinz-type means A_α and binomial means B_α (see also Chapter 1). We obtained there a variety of comparison (in terms of the order $\preceq$) among those means, which give norm inequalities including the familiar arithmetic-logarithmic-geometric mean inequality for Hilbert space operators based on Theorem 3.7 (though in [39] we restricted ourselves to the case of matrices). In the rest we will deal with the same one-parameter families of means once again, but our main aim here is to establish the norm continuity of their means of operators in the parameter α (see Theorem 5.7 for instance). In this chapter we begin with *A-L-G* interpolation means M_α while Heinz-type means A_α and binomial means B_α will be dealt with in the subsequent two chapters.

5.1 Monotonicity and related results

The most typical one-parameter family of means in $\mathfrak{M}$ is the following M_α $(-\infty \leq \alpha \leq \infty)$:

$$M_\alpha(s,t) = \begin{cases} \dfrac{\alpha - 1}{\alpha} \times \dfrac{s^\alpha - t^\alpha}{s^{\alpha-1} - t^{\alpha-1}} & (s \neq t), \\ s & (s = t), \end{cases}$$

where M_α for $\alpha = -\infty, 0, 1, \infty$ are understood as $M_{-\infty}, G, L, M_\infty$ respectively mentioned in (3.4). Indeed, notice

$$G(s,t) = \lim_{\alpha \to 0} M_\alpha(s,t), \ L(s,t) = \lim_{\alpha \to 1} M_\alpha(s,t), \ M_{\pm\infty}(s,t) = \lim_{\alpha \to \pm\infty} M_\alpha(s,t).$$

In this way, the one-parameter family M_α interpolates familiar means such as

$$M_2 = A \qquad \text{(the arithmetic mean)},$$
$$M_1 = L \qquad \text{(the logarithmic mean)},$$
$$M_{1/2} = G \qquad \text{(the geometric mean)},$$
$$M_{-1} = M_{\text{har}} \quad \text{(the harmonic mean)}.$$

The means M_α for the special values $\alpha = \frac{n}{n-1}$ $(n = 2, 3, \dots)$ and $\alpha = \frac{m}{m+1}$ $(m = 1, 2, \dots)$ are written as

$$
\begin{cases}
M_{\frac{n}{n-1}}(s,t) = \dfrac{1}{n} \times \dfrac{s^{\frac{n}{n-1}} - t^{\frac{n}{n-1}}}{s^{\frac{1}{n-1}} - t^{\frac{1}{n-1}}} = \dfrac{1}{n} \displaystyle\sum_{k=0}^{n-1} s^{\frac{k}{n-1}} t^{\frac{n-1-k}{n-1}}, \\[3ex]
M_{\frac{m}{m+1}}(s,t) = \dfrac{1}{m} \times \dfrac{s^{\frac{m}{m+1}} - t^{\frac{m}{m+1}}}{s^{\frac{-1}{m+1}} - t^{\frac{-1}{m+1}}} = \dfrac{1}{m} \displaystyle\sum_{k=1}^{m} s^{\frac{k}{m+1}} t^{\frac{m+1-k}{m+1}}.
\end{cases}
\tag{5.1}
$$

The former (resp. latter) means discretely interpolate A and L (resp. G and L), and the corresponding operator means were thoroughly investigated in [38] (where the notations A_n and G_m were used instead).

It was proved in [39] that

$$M_\alpha \preceq M_\beta \quad \text{if } -\infty \le \alpha < \beta \le \infty \tag{5.2}$$

(see (3.4) and its proof for typical cases). Hence Proposition 3.3 and Corollary 3.5 imply the following monotonicity:

Theorem 5.1. *For every $-\infty \le \alpha \le \infty$ the mean M_α is a Schur multiplier, and if $-\infty \le \alpha < \beta \le \infty$, then*

$$|||M_\alpha(H,K)X||| \le |||M_\beta(H,K)X|||$$

for all $H, K, X \in B(\mathcal{H})$ with $H, K \ge 0$ and for any unitarily invariant norm $||| \cdot |||$.

The estimate (3.16) and Theorem 5.1 guarantee the equivalence of the norms of $M_\alpha(H,K)X$ for $2 \le \alpha \le \infty$. The equivalence actually remains valid for $1 < \alpha \le \infty$ (but not for $\alpha \le 1$), as will be seen in the proposition below together with mutual norm bounds. This difference comes from the fact that $M_\alpha(1,0) > 0$ for $\alpha > 1$ in contrast with $M_\alpha(1,0) = 0$ for $\alpha \le 1$ (see Remark 5.5, (i)).

Proposition 5.2. *Let H, K be positive operators, $X \in B(\mathcal{H})$ and $||| \cdot |||$ be any unitarily invariant norm. If $1 < \alpha < \beta \le \infty$, then we have*

$$|||M_\alpha(H,K)X||| \le |||M_\beta(H,K)X|||$$
$$\le \frac{(\alpha+1)\beta - 2\alpha}{(\alpha-1)\beta} \times |||M_\alpha(H,K)X|||, \tag{5.3}$$

and

$$|||M_\alpha(H,K)X - M_\beta(H,K)X||| \leq \frac{2(\beta-\alpha)}{(\alpha-1)\beta} \times |||M_\alpha(H,K)X|||. \qquad (5.4)$$

Here, for $\beta = \infty$ the constants $\frac{(\alpha+1)\beta-2\alpha}{(\alpha-1)\beta}$ and $\frac{2(\beta-\alpha)}{(\alpha-1)\beta}$ are understood as $\frac{\alpha+1}{\alpha-1}$ and $\frac{2}{\alpha-1}$ respectively.

Proof. The first inequality in (5.3) is due to Theorem 5.1. To show the second, we first assume $1 < \alpha < \beta < \infty$. Direct computations yield

$$\frac{M_\beta(e^x,1)}{M_\alpha(e^x,1)} = \frac{\alpha(\beta-1)}{(\alpha-1)\beta} \times \frac{e^{(\alpha-1)x}-1}{e^{\alpha x}-1} \times \frac{e^{\beta x}-1}{e^{(\beta-1)x}-1}$$

$$= \frac{\alpha(\beta-1)}{(\alpha-1)\beta} \times \frac{\sinh\left(\frac{\alpha-1}{2}x\right)}{\sinh\left(\frac{\alpha}{2}x\right)} \times \frac{\sinh\left(\frac{\beta}{2}x\right)}{\sinh\left(\frac{\beta-1}{2}x\right)}. \qquad (5.5)$$

From this we easily observe

$$1 - \frac{(\alpha-1)\beta}{\alpha(\beta-1)} \times \frac{M_\beta(e^x,1)}{M_\alpha(e^x,1)}$$

$$= \frac{\sinh\left(\frac{\alpha}{2}x\right)\sinh\left(\frac{\beta-1}{2}x\right) - \sinh\left(\frac{\alpha-1}{2}x\right)\sinh\left(\frac{\beta}{2}x\right)}{\sinh\left(\frac{\alpha}{2}x\right)\sinh\left(\frac{\beta-1}{2}x\right)}$$

$$= \frac{\sinh\left(\frac{x}{2}\right)}{\sinh\left(\frac{\alpha}{2}x\right)} \times \frac{\sinh\left(\frac{\beta-\alpha}{2}x\right)}{\sinh\left(\frac{\beta-1}{2}x\right)} \qquad (5.6)$$

by using $\sinh\left(\frac{\alpha}{2}x\right) = \sinh\left(\frac{\alpha-1}{2}x\right)\cosh\left(\frac{x}{2}\right) + \cosh\left(\frac{\alpha-1}{2}x\right)\sinh\left(\frac{x}{2}\right)$ (and the similar formula for $\sinh\left(\frac{\beta}{2}x\right)$). Thanks to $1 < \alpha < \beta < \infty$, the two functions $\sinh\left(\frac{x}{2}\right)/\sinh\left(\frac{\alpha}{2}x\right)$ and $\sinh\left(\frac{\beta-\alpha}{2}x\right)/\sinh\left(\frac{\beta-1}{2}x\right)$ here are positive definite (see [39, (1.4)]) so that (5.6) is the Fourier transform of a positive measure with total mass $\frac{\beta-\alpha}{\alpha(\beta-1)}$. This means that

$$M(s,t) = \frac{\alpha(\beta-1)}{\beta-\alpha}M_\alpha(s,t) - \frac{(\alpha-1)\beta}{\beta-\alpha}M_\beta(s,t)$$

is a mean in $\mathfrak{M}$ and $M \preceq M_\alpha$ is satisfied. Therefore, from Corollary 3.5 we get

$$\left|\left|\left|\frac{\alpha(\beta-1)}{\beta-\alpha}M_\alpha(H,K)X - \frac{(\alpha-1)\beta}{\beta-\alpha}M_\beta(H,K)X\right|\right|\right| \leq |||M_\alpha(H,K)X||| \quad (5.7)$$

so that

$$\frac{(\alpha-1)\beta}{\beta-\alpha}|||M_\beta(H,K)X|||$$

$$\leq \frac{\alpha(\beta-1)}{\beta-\alpha}|||M_\alpha(H,K)X|||$$

$$\quad + \left|\left|\left|\frac{\alpha(\beta-1)}{\beta-\alpha}M_\alpha(H,K)X - \frac{(\alpha-1)\beta}{\beta-\alpha}M_\beta(H,K)X\right|\right|\right|$$

$$\leq \left(\frac{\alpha(\beta-1)}{\beta-\alpha}+1\right)|||M_\alpha(H,K)X||| = \frac{(\alpha+1)\beta-2\alpha}{\beta-\alpha} \times |||M_\alpha(H,K)X|||,$$

implying the second inequality in the case $\beta < \infty$.

The proof in the limiting case $\beta = \infty$ is similar. Indeed, we can replace the expressions (5.5) and (5.6) by

$$\frac{\alpha}{\alpha - 1} \times \frac{e^{\frac{|x|}{2}} \sinh\left(\frac{\alpha-1}{2}x\right)}{\sinh\left(\frac{\alpha}{2}x\right)} \quad \text{and} \quad e^{\frac{1-\alpha}{2}|x|} \times \frac{\sinh\left(\frac{x}{2}\right)}{\sinh\left(\frac{\alpha}{2}x\right)}$$

respectively, and then we proceed as in the above case $\beta < \infty$. Here, we point out that the function $e^{\frac{1-\alpha}{2}|x|}$ is positive definite thanks to

$$e^{-a|x|} = \frac{a}{\pi} \int_{-\infty}^{\infty} \frac{e^{ixy}}{y^2 + a^2} \, dy \qquad (a > 0) \tag{5.8}$$

(see also (7.3)).

Finally, by noting $\frac{\alpha(\beta-1)}{(\alpha-1)\beta} > 1$ and recalling (5.7), we estimate

$$|||M_\alpha(H,K)X - M_\beta(H,K)X|||$$
$$\leq \left(\frac{\alpha(\beta-1)}{(\alpha-1)\beta} - 1\right) |||M_\alpha(H,K)X|||$$
$$+ |||\frac{\alpha(\beta-1)}{(\alpha-1)\beta} M_\alpha(H,K)X - M_\beta(H,K)X|||$$
$$\leq \left(\frac{\alpha(\beta-1)}{(\alpha-1)\beta} - 1 + \frac{\beta-\alpha}{(\alpha-1)\beta}\right) |||M_\alpha(H,K)X|||.$$

The last coefficient here is $\frac{2(\beta-\alpha)}{(\alpha-1)\beta}$ so that (5.4) is obtained.   □

From the means M_α with $\alpha = \frac{n}{n-1}$ $(n = 2, 3, \dots)$ we get

$$M_{\frac{n}{n-1}}(H,K)X = \frac{1}{n} \sum_{k=0}^{n-1} H^{\frac{k}{n-1}} X K^{\frac{n-1-k}{n-1}}$$

(see (5.1)). We showed in [38] that the norm $|||\frac{1}{n} \sum_{k=0}^{n-1} H^{\frac{k}{n-1}} X K^{\frac{n-1-k}{n-1}}|||$ is monotone decreasing in n (which can be thought of as a special case of Theorem 5.1). Complementing this, we state the following special case of the above proposition:

Corollary 5.3. *Let* H, K, X *and* $|||\cdot|||$ *be as above. For all integers* $n > m \geq 2$,

$$|||\frac{1}{n} \sum_{k=0}^{n-1} H^{\frac{k}{n-1}} X K^{\frac{n-1-k}{n-1}}||| \leq |||\frac{1}{m} \sum_{k=0}^{m-1} H^{\frac{k}{m-1}} X K^{\frac{m-1-k}{m-1}}|||$$

$$\leq \frac{2n-m}{m} \times |||\frac{1}{n} \sum_{k=0}^{n-1} H^{\frac{k}{n-1}} X K^{\frac{n-1-k}{n-1}}|||$$

and

$$|||\frac{1}{n}\sum_{k=0}^{n-1} H^{\frac{k}{n-1}} XK^{\frac{n-1-k}{n-1}} - \frac{1}{m}\sum_{k=0}^{m-1} H^{\frac{k}{m-1}} XK^{\frac{m-1-k}{m-1}} |||$$

$$\leq \frac{2(n-m)}{m} \times |||\frac{1}{n}\sum_{k=0}^{n-1} H^{\frac{k}{n-1}} XK^{\frac{n-1-k}{n-1}} |||.$$

5.2 Characterization of $|||M_\infty(H,K)X||| < \infty$

The following is also a consequence of Proposition 5.2:

Proposition 5.4. *For every $H, K \geq 0$, $X \in B(\mathcal{H})$ and any unitarily invariant norm $||| \cdot |||$, the following conditions are mutually equivalent:*

(i) $|||M_\alpha(H,K)X||| < \infty$ *for some $1 < \alpha < \infty$;*
(ii) $|||M_\infty(H,K)X||| < \infty$;
(iii) $|||HX + XK||| < \infty$.

Moreover, when one (and hence all) of these conditions is satisfied, then we have the norm convergence

$$\lim_{\alpha\to\beta} |||M_\alpha(H,K)X - M_\beta(H,K)X||| = 0$$

for every $1 < \beta \leq \infty$.

Remark 5.5. A few remarks are in order.

(i)   An estimate from the above such as the second inequality in (5.3) is impossible (even for scalars) for $\alpha \leq 1$. In fact, it is straight-forward to see $\lim_{s\searrow 0} M_\beta(s,1)/M_\alpha(s,1) = \infty$ for any $\beta > \alpha$ as long as $\alpha \leq 1$.
(ii) In [53] unitarily invariant norms $||| \cdot |||$ under which the map $A \mapsto |A|$ is Lipschitz continuous were characterized as interpolation norms (see [6, 56] for general facts on interpolation spaces) between $\|\cdot\|_{p_1}$ and $\|\cdot\|_{p_2}$ with $1 < p_1, p_2 < \infty$, where the boundedness of the "upper triangular projection" played a crucial role (see [30, 59]). For such norms the inequality (5.9) in Proposition 5.6 below shows that the finiteness condition $|||HX+XK||| < \infty$ in Proposition 5.4 is equivalent to the requirement:

$$|||HX||| < \infty \quad \text{and} \quad |||XK||| < \infty.$$

Proposition 5.6. *If $||| \cdot |||$ is an interpolation norm between some Schatten p-norms $\|\cdot\|_{p_1}$ and $\|\cdot\|_{p_2}$ with $1 < p_1, p_2 < \infty$, then one can find a constant κ (depending only on $||| \cdot |||$) such that*

$$|||HX - XK||| \leq \kappa|||HX + XK||| \tag{5.9}$$

is valid for all H, K, X with $H, K \geq 0$.

Proof. The inequality (5.9) for matrices is known (see [24] and also [53]), where κ is a constant depending only upon $||| \cdot |||$ (independent of the size of matrices). We have to generalize this inequality for infinite-dimensional operators. Thanks to the standard 2×2-matrix trick we may and do assume $H = K \geq 0$. Then, for any given $\varepsilon > 0$ one can find a decomposition $H = D_\varepsilon + H_\varepsilon$ into self-adjoint operators such that $|||H_\varepsilon||| \leq \varepsilon$ and D_ε is diagonal (see [58] or [48, Chapter X, §2.2]). Note $|||H - |D_\varepsilon| \,||| \leq \text{const.} \, |||H_\varepsilon|||$ (with a constant depending only upon $||| \cdot |||$) thanks to [53, Corollary 7]. Hence, by replacing $D_\varepsilon, H_\varepsilon$ by $|D_\varepsilon|, H - |D_\varepsilon|$, we may and do assume the positivity of the diagonal operator D_ε. Notice

$$\left| \, |||D_\varepsilon X \pm X D_\varepsilon||| - |||HX \pm XH||| \, \right|$$
$$\leq |||H_\varepsilon X||| + |||X H_\varepsilon||| \leq 2|||H_\varepsilon||| \times \|X\| \leq 2\varepsilon \|X\|.$$

Thus, to show (5.9) for infinite-dimensional operators we may and do assume that $H \,(= K)$ is a positive diagonal operator from the beginning, and hence one finds a sequence $\{p_n\}_{n=1,2,\cdots}$ of finite-rank projections such that p_n tends to 1 in the strong operator topology and $H p_n = p_n H$. We then estimate

$$|||HX - XH||| \leq \liminf_{n\to\infty} |||(p_n H p_n)(p_n X p_n) - (p_n X p_n)(p_n H p_n)|||$$
$$\leq \kappa \liminf_{n\to\infty} |||(p_n H p_n)(p_n X p_n) + (p_n X p_n)(p_n H p_n)|||$$
$$\text{(by (5.9) in the matrix case)}$$
$$= \kappa \liminf_{n\to\infty} |||p_n(HX + XH)p_n|||$$
$$\leq \kappa |||HX + XH|||$$

so that (5.9) for general operators is established.   □

5.3 Norm continuity in parameter

In this section we will show the next theorem concerning the norm continuity of $M_\alpha(H, K)X$ in the parameter α.

Theorem 5.7. *Let $H, K \geq 0$, $X \in B(\mathcal{H})$ and $||| \cdot |||$ be a unitarily invariant norm. If $-\infty \leq \alpha_0 \leq \infty$ and $|||M_\beta(H, K)X||| < \infty$ for some $\beta > \min\{\alpha_0, 1\}$, then*

$$\lim_{\alpha \to \alpha_0} |||M_\alpha(H, K)X - M_{\alpha_0}(H, K)X||| = 0.$$

At first we prepare two easy lemmas for the proof of the theorem.

Lemma 5.8. *Let φ, φ_n $(n = 1, 2, \ldots)$ be nonnegative functions in $L^1(\mathbf{R})$ such that*

$$\lim_{n\to\infty} \int_{-\infty}^{\infty} \varphi_n(x)\, dx = \int_{-\infty}^{\infty} \varphi(x)\, dx.$$

If the Fourier transforms

$$\hat{\varphi}(x) = \int_{-\infty}^{\infty} e^{ixy} \varphi(y)\, dy, \quad \hat{\varphi}_n(x) = \int_{-\infty}^{\infty} e^{ixy} \varphi_n(y)\, dy$$

are in $L^2(\mathbf{R})$ *and*

$$\lim_{n\to\infty} \|\hat{\varphi}_n - \hat{\varphi}\|_2 = 0,$$

then

$$\lim_{n\to\infty} \|\varphi_n - \varphi\|_1 = 0.$$

Proof. By the Fourier inversion formula we get $\varphi, \varphi_n \in L^2(\mathbf{R})$ and

$$\|\varphi_n - \varphi\|_2 = \frac{1}{2\pi} \|\hat{\varphi}_n - \hat{\varphi}\|_2 \longrightarrow 0 \quad (n \to \infty).$$

In particular, we have the convergence $\varphi_n(x) \to \varphi(x)$ in measure. The assumption means

$$\lim_{n\to\infty} \int_{-\infty}^{\infty} (\varphi_n(x) + \varphi(x))\, dx = 2 \int_{-\infty}^{\infty} \varphi(x)\, dx.$$

Hence, by applying the extended form of the Lebesgue dominated convergence theorem (see [75, Chapter 11, Proposition 18] or [26, Theorem 3.6]) to $|\varphi_n(x) - \varphi(x)| \le \varphi_n(x) + \varphi(x)$, we conclude $\|\varphi_n - \varphi\|_1 \to 0$. $\square$

Lemma 5.9. *For any* $\theta > 0$ *and* $x > 0$ *the following inequalities hold:*

$$\text{(i)} \quad \frac{\theta}{\sinh(\theta x)} \le \frac{1}{x}, \qquad \text{(ii)} \quad \frac{\sinh(\theta x)}{\theta \sinh((1+\theta)x)} \le \frac{x}{\sinh(x)}.$$

Proof. (i) is just the well-known inequality $x \le \sinh(x)$ for $x \ge 0$. The inequality (ii) is equivalent to

$$x\theta \sinh((1+\theta)x) - \sinh(\theta x)\sinh(x) \ge 0.$$

However, it is indeed the case because the derivative (with respect to θ) of the above left-hand side is

$$x \sinh((1+\theta)x) + x^2\theta \cosh((1+\theta)x) - x\cosh(\theta x)\sinh(x)$$
$$= x\sinh(\theta x)\cosh(x) + x^2\theta \cosh((1+\theta)x) \ge 0.$$

$\square$

Proof of Theorem 5.7. The assertion for the case $1 < \alpha_0 \le \infty$ was already shown in Proposition 5.4. To deal with the case $-\infty \le \alpha_0 \le 1$, we will consider the following cases separately:

(a) $0 < \alpha_0 < 1$,   (b) $\alpha_0 < 0$,   (c) $\alpha_0 = 1$,   (d) $\alpha_0 = 0$.   (e) $\alpha_0 = -\infty$,

(a) *Case* $0 < \alpha_0 < 1$. By the assumption (also Corollary 3.5 and (5.2)) we can choose $\alpha_0 < \beta < 1$ such that $|||M_\beta(H,K)X||| < \infty$. For $0 < \alpha < \beta$ we compute

$$\frac{M_\alpha(e^x, 1)}{M_\beta(e^x, 1)} = \frac{(\alpha - 1)\beta}{\alpha(\beta - 1)} \times \frac{e^{\alpha x} - 1}{e^{(\alpha - 1)x} - 1} \times \frac{e^{(\beta - 1)x} - 1}{e^{\beta x} - 1}$$

$$= \frac{(\alpha - 1)\beta}{\alpha(\beta - 1)} \times \frac{\sinh(\frac{\alpha}{2}x)}{\sinh(\frac{\beta}{2}x)} \times \frac{\sinh(\frac{1-\beta}{2}x)}{\sinh(\frac{1-\alpha}{2}x)}$$

$$= \hat{\varphi}_{\alpha,\beta}(x)$$

for some positive function $\varphi_{\alpha,\beta} \in L^1(\mathbf{R})$ with $\int_{-\infty}^{\infty} \varphi_{\alpha,\beta}(x)dx = 1$ (see the proof of [39, Theorem 2.1] or [39, (1.4)]). We note

$$\frac{\sinh(\frac{\alpha}{2}x)}{\sinh(\frac{\beta}{2}x)} \times \frac{\sinh(\frac{1-\beta}{2}x)}{\sinh(\frac{1-\alpha}{2}x)} = O(e^{(\alpha - \beta)|x|}) \quad (\text{as } |x| \to \infty),$$

and take $\delta > 0$ satisfying $0 < \alpha_0 - \delta < \alpha_0 + \delta < \beta$. Then, $\hat{\varphi}_{\alpha,\beta}^2$ for $|\alpha - \alpha_0| < \delta$ are uniformly integrable. Moreover, it is obvious that $\hat{\varphi}_{\alpha,\beta}(x) \to \hat{\varphi}_{\alpha_0,\beta}(x)$ as $\alpha \to \alpha_0$ for all $x \in \mathbf{R}$. Thus, the Lebesgue dominated convergence theorem yields

$$\lim_{\alpha \to \alpha_0} \|\hat{\varphi}_{\alpha,\beta} - \hat{\varphi}_{\alpha_0,\beta}\|_2 = 0,$$

and so Lemma 5.8 implies

$$\lim_{\alpha \to \alpha_0} \|\varphi_{\alpha,\beta} - \varphi_{\alpha_0,\beta}\|_1 = 0.$$

Since

$$M_\alpha(H,K)X = \int_{-\infty}^{\infty} (Hs_H)^{ix}(M_\beta(H,K)X)(Ks_K)^{-ix}\varphi_{\alpha,\beta}(x)\,dx$$

$(0 < \alpha < \beta)$ by Theorem 3.4 and (5.2), we have

$$|||M_\alpha(H,K)X - M_{\alpha_0}(H,K)X||| \le \|\varphi_{\alpha,\beta} - \varphi_{\alpha_0,\beta}\|_1 \times |||M_\beta(H,K)X||| \longrightarrow 0$$

as $\alpha \to \alpha_0$.

(b) *Case* $\alpha_0 < 0$. We can choose $\alpha_0 < \beta < \frac{\alpha_0}{2}$ (or $2\beta < \alpha_0 < \beta$) such that $|||M_\beta(H,K)X||| < \infty$. When $2\beta < \alpha < \beta$, we have (see the proof of [39, Theorem 2.1])

$$\frac{M_\alpha(e^x, 1)}{M_\beta(e^x, 1)} = \frac{(1-\alpha)(-\beta)}{(-\alpha)(1-\beta)} \times \frac{\sinh(\frac{-\alpha}{2}x)}{\sinh(\frac{1-\alpha}{2}x)} \times \frac{\sinh(\frac{1-\beta}{2}x)}{\sinh(\frac{-\beta}{2}x)}$$

$$= \frac{(1-\alpha)(-\beta)}{(-\alpha)(1-\beta)} \left(1 + \frac{\sinh(\frac{x}{2})}{\sinh(\frac{1-\alpha}{2}x)} \times \frac{\sinh(\frac{\beta-\alpha}{2}x)}{\sinh(\frac{-\beta}{2}x)}\right)$$

$$= \frac{(1-\alpha)(-\beta)}{(-\alpha)(1-\beta)} + \hat{\varphi}_{\alpha,\beta}(x) \tag{5.10}$$

for some positive function $\varphi_{\alpha,\beta} \in L^1(\mathbf{R})$ with

$$\int_{-\infty}^{\infty} \varphi_{\alpha,\beta}(x) \, dx = 1 - \frac{(1-\alpha)(-\beta)}{(-\alpha)(1-\beta)}.$$

In the same way as in Case (a) we have

$$\lim_{\alpha \to \alpha_0} \|\hat{\varphi}_{\alpha,\beta} - \hat{\varphi}_{\alpha_0,\beta}\|_2 = 0$$

so that Lemma 5.8 implies $\|\varphi_{\alpha,\beta} - \varphi_{\alpha,\beta}\|_1 \to 0$ as $\alpha \to \alpha_0$. Since

$$M_\alpha(H,K)X = \int_{-\infty}^{\infty} (Hs_H)^{is} (M_\beta(H,K)X)(Ks_K)^{-ix} \varphi_{\alpha,\beta}(x) \, dx$$

$$+ \frac{(1-\alpha)(-\beta)}{(-\alpha)(1-\beta)} M_\beta(H,K)X$$

by (3.9) in Theorem 3.4, we get

$$|||M_\alpha(H,K)X - M_{\alpha_0}(H,K)X|||$$

$$\leq \left(\|\varphi_{\alpha,\beta} - \varphi_{\alpha_0,\beta}\|_1 + \left| \frac{(1-\alpha)(-\beta)}{(-\alpha)(1-\beta)} - \frac{(1-\alpha_0)(-\beta)}{(-\alpha_0)(1-\beta)} \right| \right)$$

$$\times |||M_\beta(H,K)X||| \longrightarrow 0$$

as $\alpha \to \alpha_0$.

(c) *Case $\alpha_0 = 1$.* Choose $1 < \beta < 2$ such that $|||M_\beta(H,K)X||| < \infty$. We have

$$\frac{M_1(e^x,1)}{M_\beta(e^x,1)} = \frac{\beta}{\beta-1} \times \frac{\sinh\left(\frac{x}{2}\right) \sinh\left(\frac{\beta-1}{2}x\right)}{\left(\frac{x}{2}\right) \sinh\left(\frac{\beta}{2}x\right)} = \hat{\psi}_{1,\beta}(x)$$

for some positive function $\psi_{1,\beta} \in L^1(\mathbf{R})$ with $\int_{-\infty}^{\infty} \psi_{1,\beta}(x) \, dx = 1$ by [39, Corollary 2.4]. Notice $\hat{\psi}_{1,\beta} \in L^2(\mathbf{R})$ because of $\hat{\psi}_{1,\beta}(x)^2 = O(x^{-2})$ as $|x| \to \infty$. Now we deal with the two cases $1 < \alpha < \beta$ and $0 < \alpha < 1$ separately.

First, consider the case $1 < \alpha < \beta$. We notice

$$\frac{M_\alpha(e^x,1)}{M_\beta(e^x,1)} = \frac{(\alpha-1)\beta}{\alpha(\beta-1)} \left(1 + \frac{\sinh\left(\frac{x}{2}\right) \sinh\left(\frac{\beta-\alpha}{2}x\right)}{\sinh\left(\frac{\alpha-1}{2}x\right) \sinh\left(\frac{\beta}{2}x\right)} \right)$$

$$= \frac{(\alpha-1)\beta}{\alpha(\beta-1)} + \hat{\psi}_{\alpha,\beta}(x)$$

for some positive function $\psi_{\alpha,\beta} \in L^1(\mathbf{R})$ with

$$\int_{-\infty}^{\infty} \psi_{\alpha,\beta}(x) \, dx = 1 - \frac{(\alpha-1)\beta}{\alpha(\beta-1)}.$$

Indeed, choose $\alpha = a_0 < a_1 < \cdots < a_m = \beta$ such that $a_k < 2a_{k-1} - 1$ $(1 \le k \le m)$. By the proof of [39, Theorem 2.1] there are positive functions $f_1, \ldots, f_m \in L^1(\mathbf{R})$ such that

$$\frac{M_{a_{k-1}}(e^x, 1)}{M_{a_k}(e^x, 1)} = \frac{(a_{k-1} - 1)a_k}{a_{k-1}(a_k - 1)} + \hat{f}_k(x) \qquad (1 \le k \le m)$$

so that

$$\frac{M_\alpha(e^x, 1)}{M_\beta(e^x, 1)} = \prod_{k=1}^m \left(\frac{(a_{k-1} - 1)a_k}{a_{k-1}(a_k - 1)} + \hat{f}_k(x) \right) = \frac{(\alpha - 1)\beta}{\alpha(\beta - 1)} + \hat{\psi}_{\alpha,\beta}(x).$$

Here, $\psi_{\alpha,\beta}$ is a linear combination (with positive coefficients) of the convolutions $f_{k_1} * f_{k_2} * \cdots * f_{k_l}$ for $1 \le k_1 < k_2 < \cdots < k_l \le m$ so that the positivity of $\psi_{\alpha,\beta}$ is clear.

We have

$$\hat{\psi}_{\alpha,\beta}(x) = \frac{(\alpha - 1)\beta}{\alpha(\beta - 1)} \times \frac{\sinh\left(\frac{x}{2}\right) \sinh\left(\frac{\beta - \alpha}{2} x\right)}{\sinh\left(\frac{\alpha - 1}{2} x\right) \sinh\left(\frac{\beta}{2} x\right)}$$

$$\le \frac{\beta}{\alpha(\beta - 1)} \times \frac{\sinh\left(\frac{x}{2}\right) \sinh\left(\frac{\beta - 1}{2} x\right)}{\left(\frac{x}{2}\right) \sinh\left(\frac{\beta}{2} x\right)}$$

$$= \frac{1}{\alpha} \times \hat{\psi}_{1,\beta}(x) \le \hat{\psi}_{1,\beta}(x)$$

thanks to $(\alpha - 1) / \sinh\left(\frac{\alpha-1}{2} x\right) \le \frac{2}{x}$ (see Lemma 5.9, (i)) and the increasingness $\sinh\left(\frac{\beta - \alpha}{2} x\right) \le \sinh\left(\frac{\beta - 1}{2} x\right)$ $(x \ge 0)$. Moreover, $\hat{\psi}_{\alpha,\beta}(x) \to \hat{\psi}_{1,\beta}(x)$ as $\alpha \searrow 1$ for all $x \in \mathbf{R}$. Therefore, the dominated convergence theorem shows

$$\|\hat{\psi}_{\alpha,\beta} - \hat{\psi}_{1,\beta}\|_2 = 0,$$

and Lemma 5.8 implies $\|\psi_{\alpha,\beta} - \psi_{1,\beta}\|_1 \to 0$ as $\alpha \searrow 1$. Since

$$M_1(H, K)X = \int_{-\infty}^{\infty} (Hs_H)^{ix} (M_\beta(H, K)X)(Ks_K)^{-ix} \psi_{1,\beta}(x)\, dx,$$

$$M_\alpha(H, K)X = \int_{-\infty}^{\infty} (Hs_H)^{ix} (M_\beta(H, K)X)(Ks_K)^{-ix} \psi_{\alpha,\beta}(x)\, dx$$

$$+ \frac{(\alpha - 1)\beta}{\alpha(\beta - 1)} M_\beta(H, K)X,$$

we get

$$\||M_\alpha(H, K)X - M_1(H, K)X\||$$

$$\le \left(\|\psi_{\alpha,\beta} - \psi_{1,\beta}\|_1 + \frac{(\alpha - 1)\beta}{\alpha(\beta - 1)} \right) \||M_\beta(H, K)X\|| \longrightarrow 0$$

as $\alpha \searrow 1$.

Next, consider the case $0 < \alpha < 1$. Since

$$\frac{M_\alpha(e^x, 1)}{M_\beta(e^x, 1)} = \frac{M_\alpha(e^x, 1)}{M_1(e^x, 1)} \times \hat{\psi}_{1,\beta}(x)$$

and $M_\alpha(e^x, 1)/M_1(e^x, 1)$ is a positive definite function, there is a positive function $\varphi_{\alpha,\beta} \in L^1(\mathbf{R})$ such that

$$\frac{M_\alpha(e^x, 1)}{M_\beta(e^x, 1)} = \frac{(1-\alpha)\beta}{\alpha(\beta-1)} \times \frac{\sinh\left(\frac{\alpha}{2}x\right)\sinh\left(\frac{\beta-1}{2}x\right)}{\sinh\left(\frac{1-\alpha}{2}x\right)\sinh\left(\frac{\beta}{2}x\right)} = \hat{\varphi}_{\alpha,\beta}(x).$$

We have

$$\hat{\varphi}_{\alpha,\beta}(x) \leq \frac{\beta}{\alpha(\beta-1)} \times \frac{\sinh\left(\frac{x}{2}\right)\sinh\left(\frac{\beta-1}{2}x\right)}{\left(\frac{x}{2}\right)\sinh\left(\frac{\beta}{2}x\right)} = \frac{1}{\alpha} \times \hat{\psi}_{1,\beta}(x),$$

because of $(1-\alpha)/\sinh\left(\frac{1-\alpha}{2}x\right) \leq \frac{2}{x}$ (see Lemma 5.9, (i)) and the increasingness $\sinh\left(\frac{\alpha}{2}x\right) \leq \sinh\left(\frac{x}{2}\right)$ $(x \geq 0)$. Hence we get

$$\lim_{\alpha \nearrow 1} \|\hat{\varphi}_{\alpha,\beta} - \hat{\psi}_{1,\beta}\|_2 = 0$$

by the dominated convergence theorem. Therefore, Lemma 5.8 implies $\|\varphi_{\alpha,\beta} - \psi_{1,\beta}\|_1 \to 0$ as $\alpha \nearrow 1$, and consequently

$$\lim_{\alpha \nearrow 1} |||M_\alpha(H,K)X - M_1(H,K)X||| = 0$$

as before.

(d) *Case* $\alpha_0 = 0$. Choose $0 < \beta < 1$ such that $|||M_\beta(H,K)X||| < \infty$, and deal with the two cases $0 < \alpha < \beta$ and $-\beta < \alpha < 0$ separately. For $0 \leq \alpha < \beta$ we have

$$\frac{M_\alpha(e^x, 1)}{M_\beta(e^x, 1)} = \frac{(1-\alpha)\beta}{\alpha(1-\beta)} \times \frac{\sinh\left(\frac{\alpha}{2}x\right)}{\sinh\left(\frac{\beta}{2}x\right)} \times \frac{\sinh\left(\frac{1-\beta}{2}x\right)}{\sinh\left(\frac{1-\alpha}{2}x\right)} = \hat{\varphi}_{\alpha,\beta}(x)$$

for some positive function $\varphi_{\alpha,\beta} \in L^1(\mathbf{R})$. (Here, $\frac{1}{\alpha}\sinh\left(\frac{\alpha}{2}x\right)$ for $\alpha = 0$ means $\frac{x}{2}$.) Since Lemma 5.9, (ii) gives

$$\frac{1}{\alpha} \times \sinh\left(\frac{\alpha}{2}x\right) \leq \frac{\left(\frac{x}{2}\right)\sinh\left(\frac{1+\alpha}{2}x\right)}{\sinh\left(\frac{x}{2}\right)},$$

we get

$$\hat{\varphi}_{\alpha,\beta}(x) \leq \frac{(1-\alpha)\beta}{1-\beta} \times \frac{\left(\frac{x}{2}\right)\sinh\left(\frac{1+\alpha}{2}x\right)\sinh\left(\frac{1-\beta}{2}x\right)}{\sinh\left(\frac{x}{2}\right)\sinh\left(\frac{\beta}{2}x\right)\sinh\left(\frac{1-\alpha}{2}x\right)}$$

$$\leq \frac{(1-\alpha)\beta}{1-\beta} \times \frac{\left(\frac{x}{2}\right)\sinh\left(\frac{1+(\beta/2)}{2}x\right)\sinh\left(\frac{1-\beta}{2}x\right)}{\sinh\left(\frac{x}{2}\right)\sinh\left(\frac{\beta}{2}x\right)\sinh\left(\frac{1-(\beta/2)}{2}x\right)}$$

$$\leq O(e^{-\frac{\beta}{4}|x|}) \quad (\text{as } |x| \to \infty)$$

when for example $0 < \alpha < \frac{\beta}{2}$. Hence, as usual we have $\lim_{\alpha \searrow 0} \|\hat{\varphi}_{\alpha,\beta} - \hat{\varphi}_{0,\beta}\|_2 = 0$ and so $\lim_{\alpha \searrow 0} \|\varphi_{\alpha,\beta} - \varphi_{0,\beta}\|_1 = 0$. Consequently

$$\lim_{\alpha \searrow 0} |||M_\alpha(H, K)X - M_0(H, K)X||| = 0.$$

Now, consider the case $-\beta < \alpha < 0$. We have

$$\frac{M_\alpha(e^x, 1)}{M_\beta(e^x, 1)} = \frac{(1-\alpha)\beta}{(-\alpha)(1-\beta)} \times \frac{\sinh\left(\frac{-\alpha}{2}x\right)}{\sinh\left(\frac{\beta}{2}x\right)} \times \frac{\sinh\left(\frac{1-\beta}{2}x\right)}{\sinh\left(\frac{1-\alpha}{2}x\right)} = \hat{\varphi}_{\alpha,\beta}(x)$$

for some positive $\varphi_{\alpha,\beta} \in L^1(\mathbf{R})$. Since the estimate

$$\frac{\sinh\left(\frac{-\alpha}{2}x\right)}{(-\alpha)\sinh\left(\frac{1-\alpha}{2}x\right)} \le \frac{\left(\frac{x}{2}\right)}{\sinh\left(\frac{x}{2}\right)}$$

(Lemma 5.9, (ii)) guarantees

$$\hat{\varphi}_{\alpha,\beta}(x) \le \frac{(1-\alpha)\beta}{1-\beta} \times \frac{\left(\frac{x}{2}\right)\sinh\left(\frac{1-\beta}{2}x\right)}{\sinh\left(\frac{x}{2}\right)\sinh\left(\frac{\beta}{2}x\right)} = (1-\alpha)\hat{\varphi}_{0,\beta}(x),$$

we get $\lim_{\alpha \nearrow 0} \|\hat{\varphi}_{\alpha,\beta} - \hat{\varphi}_{0,\beta}\|_2 = 0$. Therefore, we get $\lim_{\alpha \nearrow 0} \|\varphi_{\alpha,\beta} - \varphi_{0,\beta}\|_1 = 0$ and

$$\lim_{\alpha \nearrow 0} |||M_\alpha(H, K)X - M_0(H, K)X||| = 0$$

as before.

(e) *Case* $\alpha_0 = -\infty$. We may and do assume that $|||M_\beta(H, K)X||| < \infty$ for some $\beta < -1$. Then, for $\alpha < \beta$ we have $\sinh\left(\frac{\beta-\alpha}{2}x\right)/\sinh\left(\frac{1-\alpha}{2}x\right) \le 1$, and hence $\hat{\varphi}_{\alpha,\beta}(x)$ in (5.10) is majorized by the L^2-function

$$\frac{(1-\alpha)(-\beta)}{(-\alpha)(1-\beta)} \times \frac{\sinh\left(\frac{x}{2}\right)}{\sinh\left(\frac{-\beta}{2}x\right)}.$$

Therefore, it follows that $\hat{\varphi}_{\alpha,\beta}$ converges in the $\|\cdot\|_2$-norm to the function

$$\frac{-\beta}{1-\beta} \times e^{\frac{\beta-1}{2}|x|} \times \frac{\sinh\left(\frac{x}{2}\right)}{\sinh\left(\frac{-\beta}{2}x\right)}$$

as $\alpha \to -\infty$. On the other hand, we notice

$$\frac{M_{-\infty}(e^x, 1)}{M_\beta(e^x, 1)} = \frac{-\beta}{1-\beta} \times e^{-\frac{|x|}{2}} \times \frac{\sinh\left(\frac{1-\beta}{2}x\right)}{\sinh\left(\frac{-\beta}{2}x\right)}$$

$$= \frac{-\beta}{1-\beta}\left(1 + e^{\frac{\beta-1}{2}|x|} \times \frac{\sinh\left(\frac{x}{2}\right)}{\sinh\left(\frac{-\beta}{2}x\right)}\right)$$

so that the desired convergence is obtained as before.   □

For the operator norm $||| \cdot ||| = \| \cdot \|$, the boundedness requirement in Theorem 5.7 is automatic and hence we state

Corollary 5.10. *For each $H, K \geq 0$, $X \in B(\mathcal{H})$ and for each $-\infty \leq \alpha_0 \leq \infty$ we have*

$$\lim_{\alpha \to \alpha_0} \|M_\alpha(H, K)X - M_{\alpha_0}(H, K)X\| = 0.$$

In Theorem 5.7 we required the existence of $\beta > \alpha_0$ satisfying the finiteness condition $|||M_\beta(H, K)X||| < \infty$, which enabled us to combine relevant integral expressions with the Lebesgue dominated convergence theorem. We now deal with the limiting case $\beta = \alpha_0$.

Proposition 5.11. *Let $H, K \geq 0$, $X \in B(\mathcal{H})$ and $||| \cdot |||$ be a unitarily invariant norm.*

(i) *For each $-\infty \leq \alpha_0 \leq \infty$ we have*

$$\lim_{\alpha \nearrow \alpha_0} |||M_\alpha(H, K)X||| = |||M_{\alpha_0}(H, K)X||| \ (\leq \infty).$$

(ii) *Assume $-\infty < \alpha_0 \leq \infty$. If $\mathcal{I}_{|||\cdot|||}$ is uniformly convex, then as long as $M_{\alpha_0}(H, K)X \in \mathcal{I}_{|||\cdot|||}$ we have $M_\alpha(H, K)X \in \mathcal{I}_{|||\cdot|||}$ for each $\alpha \leq \alpha_0$ and the norm convergence*

$$\lim_{\alpha \nearrow \alpha_0} |||M_\alpha(H, K)X - M_{\alpha_0}(H, K)X||| = 0.$$

The result also remains valid for the ideal $\mathcal{C}_1(\mathcal{H})$ of trace class operators.

Note that the uniform convexity of $\mathcal{I}_{|||\cdot|||}$ is the same requirement as that of $\mathcal{I}_{|||\cdot|||}^{(0)}$. In fact, this condition actually implies the separability of $\mathcal{I}_{|||\cdot|||}$, i.e., $\mathcal{I}_{|||\cdot|||} = \mathcal{I}_{|||\cdot|||}^{(0)}$ (see [29, §III.6]), and that of the dual $\left(\mathcal{I}_{|||\cdot|||}\right)^*$ (see Corollary A.11 in §A.5).

Proof. (i) By the lower semi-continuity of $||| \cdot |||$ in the weak operator topology (see [37, Proposition 2.11]), Corollary 5.10 guarantees

$$|||M_{\alpha_0}(H, K)X||| \leq \liminf_{\alpha \nearrow \alpha_0} |||M_\alpha(H, K)X|||,$$

which (together with the monotonicity obtained in Theorem 5.1) shows the result.

(ii) From (i) and the uniform convexity of $\mathcal{I}_{|||\cdot|||}$, it suffices to show

$$\lim_{\alpha \nearrow \alpha_0} \phi(M_\alpha(H, K)X) = \phi(M_{\alpha_0}(H, K)X)$$

for each $\phi \in \left(\mathcal{I}_{|||\cdot|||}\right)^*$. However, thanks to the boundedness of $|||M_\alpha(H, K)X|||$ for $\alpha < \alpha_0$, we need to check this weak convergence only against ϕ's in a dense subset of $\left(\mathcal{I}_{|||\cdot|||}\right)^*$. Thanks to the separability of $\left(\mathcal{I}_{|||\cdot|||}\right)^*$, ϕ's of the form $\mathrm{Tr}(F \cdot)$ with a finite-rank operator F form a dense subspace in $\left(\mathcal{I}_{|||\cdot|||}\right)^*$. But, for $\phi = \mathrm{Tr}(F \cdot)$ the above convergence is trivial by Corollary 5.10. Finally, the assertion for $\mathcal{C}_1(\mathcal{H})$ is seen from for example [77, Theorem 2.19]. $\square$

Proposition 5.11, (ii) is meaningful only in the case $\alpha_0 \leq 1$. Actually, a situation is much better in the case $\alpha_0 > 1$; in fact, the latter case is automatically covered in Theorem 5.7. It is known ([18]) that the uniform convexity of $\mathcal{I}_{|||\cdot|||}$ is equivalent to that of the corresponding sequence Banach space. For example, the Schatten p-class $\mathcal{C}_p(\mathcal{H})$ for $1 < p < \infty$ is uniformly convex.

5.4 Notes and references

In [39] *A-L-G* interpolation means $\{M_\alpha\}_{-\infty \leq \alpha \leq \infty}$ were introduced and the monotonicity (Theorem 5.1) was proved (at least for matrices) as a refinement of the arithmetic-logarithmic-geometric mean inequality (1.8). For the special values $\alpha = \frac{n}{n-1}$ and $\alpha = \frac{m}{m+1}$ the operator means $M_{\frac{n}{n-1}}(H, K)X$ and $M_{\frac{m}{m+1}}(H, K)X$ are easy to handle for Hilbert space operators (at least as far as the definition is concerned). In fact, the expression (5.1) enables us to set

$$
\begin{cases}
M_{\frac{n}{n-1}}(H, K)X = \dfrac{1}{n} \displaystyle\sum_{k=0}^{n-1} H^{\frac{k}{n-1}} X K^{\frac{n-1-k}{n-1}}, \\[4mm]
M_{\frac{m}{m+1}}(H, K)X = \dfrac{1}{m} \displaystyle\sum_{k=1}^{m} H^{\frac{k}{m+1}} X K^{\frac{m+1-k}{m+1}}
\end{cases}
$$

directly so that detailed analysis on Schur multipliers (in Chapter 2) is irrelevant in this special case. Besides (1.8) these operator means were studied in [38]. In fact, the monotonicity of their norms (i.e., (1.9)) was shown as a refinement of (1.8).

In the appendix to [38] the norm convergence of $M_{\frac{n}{n-1}}(H, K)X$ and $M_{\frac{m}{m+1}}(H, K)X$ to the logarithmic mean $M_1(H, K)X$ was examined under suitable assumptions ([38, Propositions 6, 7, 8]). Theorem 5.7 (together with the finiteness criterion Proposition 5.4) and Proposition 5.11, (ii) in this chapter give rise to quite complete and satisfactory answers to such convergence problems to all M_α's.

Heinz-type means A_α

In this chapter we will deal with the following means in $\mathfrak{M}$:

$$A_\alpha(s,t) = A_{1-\alpha}(s,t) = \frac{1}{2}(s^\alpha t^{1-\alpha} + s^{1-\alpha}t^\alpha) \qquad (0 \leq \alpha \leq 1),$$

that interpolates the arithmetic mean $A_0 = A$ and the geometric one $A_{1/2} = G$. Obviously, each A_α is a Schur multiplier, and one has

$$A_\alpha(H,K)X = \frac{1}{2}(H^\alpha X K^{1-\alpha} + H^{1-\alpha} X K^\alpha) \qquad (6.1)$$

for all $H, K \geq 0$ and $X \in B(\mathcal{H})$ (with the convention $H^0 = K^0 = 1$ in the case $\alpha = 0, 1$). We point out that operators of this form appear in Heinz-type inequalities ([36]).

We noticed in [39] that

$$A_\alpha \preceq A_\beta \quad \text{if} \quad 0 \leq \beta < \alpha \leq \frac{1}{2}. \qquad (6.2)$$

Hence, Corollary 3.5 implies that $|||H^\alpha X K^{1-\alpha} + H^{1-\alpha} X K^\alpha|||$ is monotone decreasing in $\alpha \in [0, \frac{1}{2}]$ for unitarily invariant norms, corresponding to the well-known fact: the Heinz inequality (1.3) remains valid for these norms (see §6.3, **1**).

6.1 Norm continuity in parameter

The norm continuity of the Heinz mean $A_\alpha(H,K)X$ in the parameter α is given as follows: Let $||| \cdot |||$ be a unitarily invariant norm and $0 < \alpha_0 \leq \frac{1}{2}$. If $|||H^\beta X K^{1-\beta} + H^{1-\beta} X K^\beta||| < \infty$ for some $0 \leq \beta < \alpha_0$, then

$$\lim_{\alpha \to \alpha_0} |||(H^\alpha X K^{1-\alpha} + H^{1-\alpha} X K^\alpha) - (H^\beta X K^{1-\beta} + H^{1-\beta} X K^\beta)||| = 0.$$

This can be proved by using the integral expression

$$H^\alpha X K^{1-\alpha} + H^{1-\alpha} X K^\alpha$$
$$= \int_{-\infty}^{\infty} (H s_H)^{ix} (H^\beta X K^{1-\beta} + H^{1-\beta} X K^\beta)(K s_K)^{-ix} f_{\alpha,\beta}(x)\, dx$$

for $0 \leq \beta < \alpha \leq \frac{1}{2}$, where $f_{\alpha,\beta}$ is a positive function with $\int_{-\infty}^{\infty} f_{\alpha,\beta}(x)\, dx = 1$ such that

$$\frac{A_\alpha(e^x, 1)}{A_\beta(e^x, 1)} = \frac{\cosh((\frac{1}{2} - \alpha)x)}{\cosh((\frac{1}{2} - \beta)x)} = \hat{f}_{\alpha,\beta}(x).$$

In fact, we have an explicit form of the function $f_{\alpha,\beta}$ (see [39, (1.5)]), and so the proof is much easier than that of Theorem 5.7. Moreover, the above norm convergence can be improved in Proposition 6.1 below.

Note that the convergence $A_\alpha(H, K)X \to A(H, K)X$ as $\alpha \to 0$ is not true even in the matrix case. In fact, when P, Q are orthogonal projections with $P \perp Q$ and $X = 1$, we have $A(P, Q)1 = \frac{1}{2}(P + Q)$ but $A_\alpha(P, Q)1 = 0$ for all $0 < \alpha \leq \frac{1}{2}$.

One piece $H^\alpha X K^{1-\alpha}$ of the mean (6.1) is asymmetric, however our method using integral expressions can still work to treat it. Actually, the following integral formula was obtained in [54, Theorem 6]:

$$H^\alpha X K^{1-\alpha} = \int_{-\infty}^{\infty} (H s_H)^{ix} (HX + XK)(K s_K)^{-ix}$$
$$\times \frac{dx}{2\cosh(\pi x + \pi i(\alpha - \frac{1}{2}))} \tag{6.3}$$

for each $0 < \alpha < 1$ and for all $H, K, X \in B(\mathcal{H})$ with $H, K \geq 0$. (A particular case of this was given in Example 3.6, (a).) Let $g_\alpha(x)$ be the density appearing in (6.3). Then, Theorem A.5 implies

$$|||H^\alpha X K^{1-\alpha}||| \leq \left(\int_{-\infty}^{\infty} |g_\alpha(x)|\, dx \right) |||HX + XK|||, \tag{6.4}$$

which is the weak matrix Young inequality in [54] (see also [3]).

Proposition 6.1. *Let $H, K \geq 0$, $X \in B(\mathcal{H})$ and $|||\cdot|||$ be a unitarily invariant norm. If $0 < \alpha_0 < 1$ and $|||H^\beta X K^{1-\beta} + H^\gamma X K^{1-\gamma}||| < \infty$ for some $0 \leq \gamma < \alpha_0 < \beta \leq 1$ (this is the case in particular when $|||HX + XK||| < \infty$), then*

$$\lim_{\alpha \to \alpha_0} |||H^\alpha X K^{1-\alpha} - H^{\alpha_0} X K^{1-\alpha_0}||| = 0.$$

Proof. Since

$$H^\beta X K^{1-\beta} + H^\gamma X K^{1-\gamma} = H^{\beta-\gamma}(H^\gamma X K^{1-\beta}) + (H^\gamma X K^{1-\beta})K^{\beta-\gamma}$$

and

$$H^\alpha X K^{1-\alpha} = (H^{\beta-\gamma})^{\frac{\alpha-\gamma}{\beta-\gamma}} (H^\gamma X K^{1-\beta})(K^{\beta-\gamma})^{1-\frac{\alpha-\gamma}{\beta-\gamma}},$$

we may and do assume $|||HX+XK||| < \infty$ (i.e., $\beta = 1$ and $\gamma = 0$) by replacing H, K, X by $H^{\beta-\gamma}, K^{\beta-\gamma}, H^{\gamma}XK^{1-\beta}$ respectively. Then by (6.3) and Theorem A.5, we get

$$|||H^{\alpha}XK^{1-\alpha} - H^{\alpha_0}XK^{1-\alpha_0}||| \le \|g_{\alpha} - g_{\alpha_0}\|_1 \times |||HX + XK|||$$

for $0 < \alpha, \alpha_0 < 1$. Thus, it suffices to see that $\|g_{\alpha} - g_{\alpha_0}\|_1 \to 0$ as $\alpha \to \alpha_0$. However, by recalling ([54, p. 443])

$$|g_{\alpha}(x)| = \frac{1}{2\sqrt{\sinh^2(\pi x) + \cos^2(\pi(\alpha - \frac{1}{2}))}}, \tag{6.5}$$

we see that the above L^1-convergence is an immediate consequence of the Lebesgue dominated convergence theorem. □

6.2 Convergence of operator Riemann sums

We present another application of the integral expression (6.3) in a similar nature. Let us consider the following operator Riemann sum:

$$R(n) = \frac{1}{n}\sum_{k=1}^{n} H^{\xi_k}XK^{1-\xi_k} \quad (\text{with } \xi_k \in [\tfrac{k-1}{n}, \tfrac{k}{n}]).$$

From (6.3) we get

$$R(n) = \int_{-\infty}^{\infty} (Hs_H)^{ix}(HX + XK)(Ks_K)^{-ix}\phi_n(x)\,dx$$

with

$$\phi_n(x) = \frac{1}{n}\sum_{k=1}^{n} \frac{1}{2\cosh(\pi x + \pi i(\xi_k - \frac{1}{2}))}.$$

For a moment we assume that Riemann sums are chosen symmetrically, i.e., $\xi_{n+1-k} = 1 - \xi_k$ for each n and k. (Asymmetric Riemann sums will be considered in Proposition 6.3.) Then, we easily compute

$$\begin{aligned}
\phi_{2m}(x) &= \frac{1}{2m}\sum_{k=1}^{2m} \frac{1}{2\cosh(\pi x + \pi i(\xi_k - \frac{1}{2}))} \\
&= \frac{1}{2m}\sum_{k=1}^{m} \frac{\cosh(\pi x)\cos(\pi(\xi_k - \frac{1}{2}))}{\cos^2(\pi(\xi_k - \frac{1}{2})) + \sinh^2(\pi x)}
\end{aligned} \tag{6.6}$$

thanks to

$$\cosh(\pi x + \pi i(\xi_k - \tfrac{1}{2}))$$
$$= \cosh(\pi x)\cos(\pi(\xi_k - \tfrac{1}{2})) + i\sinh(\pi x)\sin(\pi(\xi_k - \tfrac{1}{2})).$$

We similarly get

$$\phi_{2m+1}(x)$$
$$= \frac{1}{2m+1} \left(\sum_{k=1}^{m} \frac{\cosh(\pi x) \cos\left(\pi\left(\xi_k - \frac{1}{2}\right)\right)}{\cos^2\left(\pi\left(\xi_k - \frac{1}{2}\right)\right) + \sinh^2(\pi x)} + \frac{1}{2\cosh(\pi x)} \right), \quad (6.7)$$

where the last term arises from the midpoint $\xi_{m+1} = \frac{1}{2}$. On the other hand, the logarithmic mean is given by

$$L = \int_0^1 H^s X K^{1-s} ds = \int_{-\infty}^{\infty} (H s_H)^{ix} (HX + XK)(K s_K)^{-ix} \phi(x) \, dx$$

with

$$\phi(x) = \frac{1}{\pi} \log \left| \coth\left(\frac{\pi x}{2}\right) \right| \quad (6.8)$$

(see [38, p. 305]).

Proposition 6.2. *Let $H, K \geq 0$, $X \in B(\mathcal{H})$, and we assume $|||HX+HK||| < \infty$ for a unitarily invariant norm $|||\cdot|||$. Then, as long as Riemann sums $R(n)$ are chosen symmetrically (i.e., $\xi_{n+1-k} = 1 - \xi_k$ for each n and k) we have*

$$\lim_{n \to \infty} |||R(n) - \int_0^1 H^s X K^{1-s} ds||| = 0.$$

Proof. From the preceding integral expressions for $R(n)$ and L we see

$$R(n) - L = \int_{-\infty}^{\infty} (H s_H)^{ix} (HX + XK)(K s_K)^{-ix} (\phi_n(x) - \phi(x)) \, dx$$

so that Theorem A.5 shows

$$|||R(n) - L||| \leq \|\phi_n - \phi\|_1 \times |||HX + XK|||.$$

Hence, as usual it suffices to see $\lim_{n\to\infty} \|\phi_n - \phi\|_1 = 0$. The Fourier transform of the positive and positive definite function ϕ is $\frac{1}{x} \tanh\left(\frac{x}{2}\right)$ (see [38, p. 306]) and hence

$$\int_{-\infty}^{\infty} \phi(x) \, dx = \frac{1}{2}.$$

On the other hand, the positive (and actually positive definite) function ϕ_n also satisfies

$$\int_{-\infty}^{\infty} \phi_n(x) \, dx = \frac{1}{2}$$

because of (6.6), (6.7) and

$$\int_{-\infty}^{\infty} \frac{\cosh(\pi x)}{\cos^2\left(\pi\left(\xi_k - \frac{1}{2}\right)\right) + \sinh^2(\pi x)} \, dx = \frac{1}{\cos\left(\pi\left(\xi_k - \frac{1}{2}\right)\right)},$$

$$\int_{-\infty}^{\infty} \frac{1}{\cosh(\pi x)} \, dx = 1.$$

(The fact $\int_{-\infty}^{\infty} \phi(x)dx = \int_{-\infty}^{\infty} \phi_n(x)dx = \frac{1}{2}$ can be also seen by simply setting $H = K = X = 1$ in the integral expressions for $R(n)$ and L.) We claim

$$\lim_{n \to \infty} \phi_n(x) = \phi(x).$$

Indeed, from (6.6) and (6.7) we observe that the limit in the left-hand side is equal to the following definite integral:

$$\cosh(\pi x) \int_{-\frac{1}{2}}^{0} \frac{\cos(\pi\alpha)}{\cos^2(\pi\alpha) + \sinh^2(\pi x)} \, d\alpha$$

$$= \cosh(\pi x) \int_{-\frac{1}{2}}^{0} \frac{\cos(\pi\alpha)}{\cosh^2(\pi x) - \sin^2(\pi\alpha)} \, d\alpha$$

$$= \frac{\cosh(\pi x)}{\pi} \int_{-1}^{0} \frac{1}{\cosh^2(\pi x) - t^2} \, dt$$

$$= \frac{1}{2\pi} \int_{-1}^{0} \left(\frac{1}{\cosh(\pi x) + t} + \frac{1}{\cosh(\pi x) - t} \right) dt$$

$$= -\frac{1}{2\pi} \log \left| \frac{\cosh(\pi x) - 1}{\cosh(\pi x) + 1} \right|,$$

which is obviously $\phi(x)$. The desired L^1-convergence thus follows from the extended Lebesgue dominated convergence theorem (see [75, Chapter 11, Proposition 18]). $\square$

Proposition 6.2 (as well as Theorem 5.7) is a considerable generalization of the convergence results obtained in the appendix to [38]. If Riemann sums are asymmetric, then $|||R(n)||| < \infty$ is no longer guaranteed (under the assumption $|||HX + XK||| < \infty$). However, for interpolation norms between $\|\cdot\|_{p_1}$ and $\|\cdot\|_{p_2}$ with $1 < p_1, p_2 < \infty$ (see Remark 5.5, (ii)), the finiteness $|||R(n)||| < \infty$ is indeed guaranteed (see the inequality at the beginning of the proof of the proposition below). Actually, for such norms we have the following strengthening of Proposition 6.2:

Proposition 6.3. *For an interpolation norm between* $\|\cdot\|_{p_1}$ *and* $\|\cdot\|_{p_2}$ *with* $1 < p_1, p_2 < \infty$ *the convergence in Proposition 6.2 remains valid for general Riemann sums (which are not necessarily symmetric).*

Proof. We choose and fix a small $\varepsilon > 0$, and split the sum $\sum_{k=1}^{n} H^{\xi_k} X K^{1-\xi_k}$ (appearing in the definition of the Riemann sum $R(n)$) into the following two parts:

$$\begin{cases} \sum' : \text{summation over } k\text{'s satisfying } [\frac{k-1}{n}, \frac{k}{n}] \subseteq [\varepsilon, 1 - \varepsilon], \\ \sum'' : \text{summation over other } k\text{'s.} \end{cases}$$

Thanks to the assumption on the norm $||| \cdot |||$, we have

$$|||H^\alpha X K^{1-\alpha}||| \leq \kappa |||HX + XK||| \quad \text{(for each } \alpha \in [0,1]\text{)}$$

with a constant κ (depending only on $||| \cdot |||$) (see [39, Proposition 3.1] and Proposition 5.6). By counting the number of subintervals "near the end-points", this inequality guarantees

$$||| \frac{1}{n} \sum {}'' H^{\xi_k} X K^{1-\xi_k} ||| \le \frac{2(n\varepsilon + 1)}{n} \times \kappa |||HX + XK|||$$

$$= 2\left(\varepsilon + \frac{1}{n}\right) \kappa |||HX + XK|||.$$

We similarly get

$$||| \int_0^\varepsilon H^s X K^{1-s} ds + \int_{1-\varepsilon}^1 H^s X K^{1-s} ds ||| \le 2\varepsilon\kappa |||HX + XK|||.$$

From the estimates so far (near the endpoints), we conclude

$$||| R(n) - \int_0^1 H^s X K^{1-s} ds |||$$

$$\le ||| \frac{1}{n} \sum {}' H^{\xi_k} X K^{1-\xi_k} - \int_\varepsilon^{1-\varepsilon} H^s X K^{1-s} ds |||$$

$$+ 2\left(2\varepsilon + \frac{1}{n}\right) \kappa |||HX + XK|||. \tag{6.9}$$

To see the limit (as $n \to \infty$) of the first quantity in the right-hand side of (6.9), we need to check the behavior of Riemann sums corresponding to the interval $[\varepsilon, 1 - \varepsilon]$. From (6.3) we get

$$\frac{1}{n} \sum {}' H^{\xi_k} X K^{1-\xi_k} = \int_{-\infty}^\infty (Hs_H)^{ix} (HX + XK)(Ks_K)^{-ix} \psi_n(x) \, dx,$$

$$\int_\varepsilon^{1-\varepsilon} H^s X K^{1-s} ds = \int_{-\infty}^\infty (Hs_H)^{ix} (HX + XK)(Ks_K)^{-ix} \psi(x) \, dx$$

with the densities

$$\psi_n(x) = \frac{1}{n} \sum {}' \frac{1}{2\cosh(\pi x + \pi i(\xi_k - \frac{1}{2}))},$$

$$\psi(x) = \int_\varepsilon^{1-\varepsilon} \frac{1}{2\cosh(\pi x + \pi i(\alpha - \frac{1}{2}))} \, d\alpha.$$

Of course we have $\lim_{n\to\infty} \psi_n(x) = \psi(x)$ from the definition of $\sum'$ and the continuity of the involved function. On the other hand, from (6.5) we observe

$$|\psi_n(x)| \le \frac{1}{2n} \sum {}' \frac{1}{\sqrt{\sinh^2(\pi x) + \cos^2(\pi(\xi_k - \frac{1}{2}))}}.$$

From the definition of $\sum'$ we have $\varepsilon \le \xi_k \le 1 - \varepsilon$ and hence

$$\cos\left(\pi\left(\xi_k - \tfrac{1}{2}\right)\right) \geq \cos\left(\pi\left(\tfrac{1}{2} - \varepsilon\right)\right) > 0,$$

which enables us to obtain the following uniform (independent of n) bound:

$$|\psi_n(x)| \leq \frac{1}{2}\min\left\{\frac{1}{|\sinh(\pi x)|}, \frac{1}{\cos\left(\pi\left(\tfrac{1}{2} - \varepsilon\right)\right)}\right\}.$$

The right side here being an L^1-function, we see $\lim_{n\to\infty}\|\psi_n - \psi\|_1 = 0$ by the Lebesgue dominated convergence theorem. The usual argument thus shows

$$\left|\left|\left|\frac{1}{n}\sum{}'H^{\xi_k}XK^{1-\xi_k} - \int_\varepsilon^{1-\varepsilon} H^s XK^{1-s}ds\right|\right|\right|$$
$$\leq \|\psi_n - \psi\|_1 \times |||HX + XK||| \to 0$$

as $n \to \infty$. Therefore, (6.9) implies

$$\limsup_{n\to\infty}\left|\left|\left|R(n) - \int_0^1 H^s XK^{1-s}ds\right|\right|\right| \leq 4\varepsilon\kappa|||HX + XK|||,$$

and consequently we get

$$\lim_{n\to\infty}\left|\left|\left|R(n) - \int_0^1 H^s XK^{1-s}ds\right|\right|\right| = 0$$

due to the arbitrariness of $\varepsilon > 0$. $\square$

6.3 Notes and references

1. Heinz inequality

The Heinz inequality (1.3) (in the operator norm) is equivalent to the decreasingness of the function

$$\alpha \in [0, 1/2] \mapsto \|H^\alpha XK^{1-\alpha} + H^{1-\alpha}XK^\alpha\|.$$

For the special value $\alpha = \tfrac{1}{2}$ the Heinz inequality reduces to the arithmetic-geometric inequality (1.4) (in the operator norm). The original proof in [36] was quite involved, and in [64] A. McIntosh presented a simpler proof in two steps: (i) a direct proof of the latter is obtained, (ii) the former is proved from the latter by certain iteration arguments. In [10] the latter was shown to remain valid for unitarily invariant norms, and hence so does the former (i.e., the Heinz inequality). In fact, (although quite ingenious) the step (ii) is based on just the triangle inequality (see [64, Theorem 4]). A slightly different proof can be found in [63, Theorem 2.3]. Proofs can be also found in [13, 54]. The proof in [13] uses a Schur multiplier while that in [54, 39] uses an integral formula of the form (3.9) (which arises from the Poisson integral formula below). Both proofs are essentially based on the positive definiteness of

$$\frac{\cosh(ax)}{\cosh(x)} = \int_{-\infty}^{\infty} \frac{\cos(\pi a/2)\cosh(\pi y/2)}{\cosh(\pi y) + \cos(\pi a)} e^{ixy} dy \quad (0 \le a < 1).$$

On the other hand,

$$\frac{\sinh(ax)}{\sinh(x)} = \frac{1}{2}\int_{-\infty}^{\infty} \frac{\sin(\pi a)}{\cosh(\pi y) + \cos(\pi a)} e^{ixy} dy \quad (0 < a < 1)$$

is also positive definite, which corresponds to the difference version

$$|||H^\theta X K^{1-\theta} - H^{1-\theta} X K^\theta||| \le |2\theta - 1| \times |||HX - XK||| \quad (\text{for } \theta \in [0,1])$$

of the Heinz inequality (see [13, p. 219] or [54, p. 435] for instance). From this we get the following inequality (see [53, Theorem 4]):

$$|||HX - XK||| \le |||e^{H/2} X e^{-K/2} - e^{-H/2} X e^{K/2}|||,$$

where H, K are self-adjoint operators. This commutator estimate also follows from the positive definiteness of the function $x/\sinh(x)$ (see (3.5)).

2. Matrix Young inequality and related topics

Almost all results in this chapter are based on the integral expression (6.3). This formula appeared in [54], from which the weak Young inequality ((1.6) and (6.4)) was derived. This inequality was motivated by T. Ando's work [3] on the (operator) Young inequality (1.5). The special case $p = q = 2$ was obtained earlier by R. Bhatia and F. Kittaneh ([12]). Note that (1.5) actually implies

$$|||f(|H^{\frac{1}{p}} K^{\frac{1}{q}}|)||| \le |||f(\tfrac{1}{p}H + \tfrac{1}{q}K)|||$$

for $p, q > 1$ with $p^{-1} + q^{-1} = 1$ and a continuous increasing function f on $[0, \infty)$ satisfying $f(0) = 0$.

We observed

$$H^{\frac{1}{p}} X K^{\frac{1}{q}} = \int_{-\infty}^{\infty} H^{ix}(HX) K^{-ix} \frac{\sin(\pi/q)}{2(\cosh(\pi x) - \cos(\pi/q))} dx$$
$$+ \int_{-\infty}^{\infty} H^{ix}(XK) K^{-ix} \frac{\sin(\pi/q)}{2(\cosh(\pi x) + \cos(\pi/q))} dx$$

in [54, §2]. This is nothing but the Poisson integral formula (for the strip $0 \le \operatorname{Im} z \le 1$) applied for $f(z) = H^{-iz} X K^{1+iz}$, and the reason why the Fourier transform of $\sin(ax)/\sin(x)$ is given as above was also explained in [54, Appendix B]. This integral expression immediately yields (1.7). We point out that (1.7) is actually equivalent to the following multiplicative version:

$$|||H^{\frac{1}{p}} X K^{\frac{1}{q}}||| \le |||HX|||^{1/p} |||XK|||^{1/q}.$$

Indeed, (1.7) comes from the multiplicative version together with the Young inequality (for scalars). On the other hand, with $t^p H$ and K/t^q $(t > 0)$ instead of H, K (1.7) gives us $|||H^{\frac{1}{p}}XK^{\frac{1}{q}}||| \leq \frac{t^p}{p}|||HX||| + \frac{t^{-q}}{q}|||XK|||$. The minimum of the right side here is $|||HX|||^{1/p}|||XK|||^{1/q}$ as desired. The multiplicative version first appeared in [11] by R. Bhatia and C. Davis (see also [64, Theorem 4, (iii)]). It can be further extended for example to

$$||| \, |H^{\frac{1}{p}}XK^{\frac{1}{q}}|^r \, ||| \leq ||| \, |HX|^r|||^{1/p}||| \, |XK|^r|||^{1/q}$$
$$\left(\leq \frac{1}{p}||| \, |HX|^r||| + \frac{1}{q}||| \, |XK|^r||| \right)$$

with $r > 0$ (see [43, Theorem 3], [54, Theorem 3] for instance). An updated survey on these Hölder-type norm inequalities can be found in [84, §4.4].

7

Binomial means B_α

The "binomial means" introduced in [39] are

$$B_\alpha(s,t) = \left(\frac{s^\alpha + t^\alpha}{2}\right)^{1/\alpha} \qquad (-\infty \le \alpha \le \infty).$$

For special values of α we have

$$\begin{aligned} B_1 &= A && \text{(the arithmetic mean)} \\ B_0 &= G && \text{(the geometric mean)} \\ B_\infty &= M_\infty. \end{aligned}$$

In fact, notice $\lim_{\alpha \to 0} B_\alpha(s,t) = G(s,t)$ and $\lim_{\alpha \to \pm\infty} B_\alpha(s,t) = M_{\pm\infty}(s,t)$. In this chapter we will prove that the binomial means are Schur multipliers, and norm continuity (in parameter) will be also discussed.

7.1 Majorization $B_\alpha \preceq M_\infty$

For means M $(= M_\alpha, A_\alpha)$ in the preceding chapters the majorization $M \preceq M_\infty$ (which ensures that M is a Schur multiplier) is relatively easy to establish. We also have $B_\alpha \preceq M_\infty$, however more involved arguments are needed.

In what follows we (mainly) assume $\alpha > 0$ and $\alpha \ne \frac{1}{n}$ $(n = 1, 2, \ldots)$. It is plain to see

$$\frac{B_\alpha(e^x, 1)}{M_\infty(e^x, 1)} = \left(\frac{1 + e^{-\alpha|x|}}{2}\right)^{1/\alpha} \tag{7.1}$$

so that we have

$$\frac{B_\alpha(e^x, 1)}{M_\infty(e^x, 1)} - 2^{-1/\alpha} = 2^{-1/\alpha}\left(\left(1 + e^{-\alpha|x|}\right)^{1/\alpha} - 1\right).$$

We set

$$\phi_\alpha(x) = \left(1 + e^{-\alpha|x|}\right)^{1/\alpha} - 1 = \left(1 + e^{-\alpha|x|}\right)^\beta - 1 \tag{7.2}$$

with $\beta = 1/\alpha \in \mathbf{R}_+ \setminus \mathbf{N}$.

We consider the power series expansion of the analytic function

$$f_\beta(z) = (1 - z)^\beta.$$

The radius of convergence here is obviously 1, and the n-th coefficient is given by

$$a_n = \frac{(-1)^n}{n!} \times \beta(\beta - 1)(\beta - 2)\cdots(\beta - (n-1))$$

for $n = 1, 2, \ldots$ and $a_0 = 1$.

The next lemma is an obvious extension of the one presented in [74, p. 195], which will be repeatedly used.

Lemma 7.1. *We have the absolute convergence* $\sum_{n=0}^\infty |a_n| < \infty$.

Proof. From the above expression we observe that a_n's are either all negative or all positive (depending upon the parity of n_0) for each $n \geq n_0 = [\beta] + 1$.

We first assume $a_n < 0$ for $n \geq n_0$. For a real t with $0 < t < 1$ we have

$$(1 - t)^\beta - \sum_{n=0}^{n_0-1} a_n t^n = \sum_{n=n_0}^N a_n t^n + \sum_{n=N+1}^\infty a_n t^n \leq \sum_{n=n_0}^N a_n t^n$$

for N large enough. Therefore, we have

$$\sum_{n=n_0}^N |a_n| = -\sum_{n=n_0}^N a_n = -\lim_{t \nearrow 1} \sum_{n=n_0}^N a_n t^n$$

$$\leq \lim_{t \nearrow 1} \left(\sum_{n=0}^{n_0-1} a_n t^n - (1-t)^\beta \right) = \sum_{n=0}^{n_0-1} a_n.$$

By letting $N \to \infty$, we see

$$\sum_{n=n_0}^\infty |a_n| \leq \sum_{n=0}^{n_0-1} a_n.$$

We next assume $a_n > 0$ for $n \geq n_0$ so that we have the reversed inequality

$$(1 - t)^\beta - \sum_{n=0}^{n_0-1} a_n t^n \geq \sum_{n=n_0}^N a_n t^n$$

for $0 < t < 1$. In this case we estimate

$$\sum_{n=n_0}^{N} |a_n| = \sum_{n=n_0}^{N} a_n = \lim_{t \nearrow 1} \sum_{n=n_0}^{N} a_n t^n$$

$$\leq \lim_{t \nearrow 1} \left((1-t)^\beta - \sum_{n=0}^{n_0-1} a_n t^n \right) = - \sum_{n=0}^{n_0-1} a_n,$$

and hence

$$\sum_{n=n_0}^{\infty} |a_n| \leq - \sum_{n=0}^{n_0-1} a_n$$

by letting $N \to \infty$ again. $\quad\square$

By substituting $z = -e^{-\alpha|x|} \in [-1,0)$ to $f_\beta(z)$ and then subtracting 1, we have

$$\phi_\alpha(x) = \sum_{n=0}^{\infty} a_n (-e^{-\alpha|x|})^n - 1 = \sum_{n=1}^{\infty} (-1)^n a_n e^{-n\alpha|x|} = \sum_{n=1}^{\infty} b_n e^{-n\alpha|x|}$$

(see (7.2)) with

$$b_n = (-1)^n a_n = \frac{1}{n!} \times \beta(\beta-1)(\beta-2)\cdots(\beta-(n-1))$$

for $n \geq 1$. Note $b_n > 0$ up to $n = n_0$ and then the signs of b_n's oscillates (i.e., $b_{n_0+1}, b_{n_0+3}, \cdots < 0$). The above expression of $\phi_\alpha(t)$ is absolutely convergent thanks to Lemma 7.1, which guarantees the validity of the following re-grouping of terms:

$$\phi_\alpha(x) = \phi_{\alpha,+}(x) - \phi_{\alpha,-}(x)$$

with

$$\begin{cases} \phi_{\alpha,+}(x) = \sum_{b_n>0} b_n e^{-n\alpha|x|} = \sum_{n=1}^{n_0-1} b_n e^{-n\alpha|x|} + \sum_{n=0}^{\infty} b_{n_0+2n} e^{-(n_0+2n)\alpha|x|}, \\[3mm] \phi_{\alpha,-}(x) = \sum_{b_n<0} (-b_n) e^{-n\alpha|x|} = \sum_{n=0}^{\infty} (-b_{n_0+2n+1}) e^{-(n_0+2n+1)\alpha|x|} \end{cases}$$

(with the convention $\sum_{n=1}^{n_0-1} = 0$ in the case $n_0 = 1$).

Theorem 7.2. *For each $\alpha \in [-\infty, \infty]$ we have $B_\alpha \preceq M_\infty$.*

Proof. Recall $B_{\pm\infty} = M_{\pm\infty}$ and $B_0 = G = M_{1/2}$ (the geometric mean), for which the result is known (Theorem 5.1). For $\alpha = 1/n$ we obviously have $B_\alpha \preceq M_\infty$ because the binomial expansion is available. We also recall $M_{1/2} \preceq B_{-\alpha}$ for $\alpha < 0$ ([39, Proposition 3.3]) so that we observe

$$B_\alpha = B^{(-)}_{-\alpha} \preceq M^{(-)}_{1/2} = M_{1/2} \preceq M_\infty$$

(see (3.18)). Therefore, in the rest of the proof we may and do assume $\alpha > 0$ and $\alpha \neq \frac{1}{n}$ ($n = 1, 2, \cdots$), and it suffices to see that the function $\phi_\alpha(t)$ (see (7.2)) is positive definite.

Recall that the Fourier transform of $e^{-a|x|}$ (with $a > 0$) is $2a(x^2 + a^2)^{-1}$:

$$\int_{-\infty}^{\infty} e^{-a|y|} e^{ixy} dy = 2\pi \times \frac{a}{\pi} \times \frac{1}{x^2 + a^2} = \frac{2a}{x^2 + a^2} \tag{7.3}$$

(thanks to (5.8) or by elementary direct computations). Lemma 7.1 and the obvious estimate $e^{-n\alpha|x|} \leq e^{-\alpha|x|}$ enable us to perform term-wise Fourier transform for the above $\phi_{\alpha,\pm}(x)$ (thanks to the dominated convergence theorem), and we get

$$\begin{cases} \hat{\phi}_{\alpha,+}(x) = \sum_{n=1}^{n_0-1} b_n \dfrac{2n\alpha}{x^2 + (n\alpha)^2} + \sum_{n=0}^{\infty} b_{n_0+2n} \dfrac{2(n_0 + 2n)\alpha}{x^2 + (n_0 + 2n)^2\alpha^2}, \\[3mm] \hat{\phi}_{\alpha,-}(x) = \sum_{n=0}^{\infty} (-b_{n_0+2n+1}) \dfrac{2(n_0 + 2n + 1)\alpha}{x^2 + (n_0 + 2n + 1)^2\alpha^2} \end{cases}$$

due to (7.3). To establish the positivity of $\hat{\phi}_\alpha(x) = \hat{\phi}_{\alpha,+}(x) - \hat{\phi}_{\alpha,-}(x)$ (i.e., the positive definiteness of ϕ_α), we will make use of the expression

$$\hat{\phi}_\alpha(x) - 2 \sum_{n=1}^{n_0-1} b_n \frac{n\alpha}{x^2 + (n\alpha)^2}$$

$$= 2 \sum_{n=0}^{\infty} \left(b_{n_0+2n} \frac{(n_0 + 2n)\alpha}{x^2 + (n_0 + 2n)^2\alpha^2} \right.$$

$$\left. -(-b_{n_0+2n+1}) \frac{(n_0 + 2n + 1)\alpha}{x^2 + (n_0 + 2n + 1)^2\alpha^2} \right).$$

Indeed, because of $\frac{n\alpha}{x^2 + n^2\alpha^2} \leq \frac{1}{n\alpha} \leq \frac{1}{\alpha}$ and Lemma 7.1, the above sums for $\hat{\phi}_{\alpha,\pm}(x)$ are once again absolutely convergent so that re-grouping terms is certainly legitimate. We note

$$\frac{-b_{n_0+2n+1}}{b_{n_0+2n}} = \frac{-(\beta - (n_0 + 2n))}{n_0 + 2n + 1} = \frac{n_0 + 2n - \beta}{n_0 + 2n + 1} = \frac{2n + \gamma}{n_0 + 2n + 1}$$

with $\gamma = n_0 - \beta = [\beta] + 1 - \beta \in (0, 1)$. Therefore, we can rewrite the above quantity as follows:

$$\hat{\phi}_\alpha(x) - 2 \sum_{n=1}^{n_0-1} b_n \frac{n\alpha}{x^2 + (n\alpha)^2}$$

$$= 2 \sum_{n=0}^{\infty} b_{n_0+2n} \left(\frac{(n_0 + 2n)\alpha}{x^2 + (n_0 + 2n)^2\alpha^2} \right.$$

$$\left. - \frac{2n+\gamma}{n_0 + 2n + 1} \times \frac{(n_0 + 2n + 1)\alpha}{x^2 + (n_0 + 2n + 1)^2\alpha^2} \right)$$

$$= 2 \sum_{n=0}^{\infty} \alpha b_{n_0+2n} \left(\frac{n_0 + 2n}{x^2 + (n_0 + 2n)^2\alpha^2} - \frac{2n+\gamma}{x^2 + (n_0 + 2n + 1)^2\alpha^2} \right).$$

Hence, it suffices to check that the difference appearing in the above last parenthesis is positive. However, by elementary computation this quantity is equal to

$$\frac{(n_0 - \gamma)x^2 + (n_0 + 2n)\left(1 + (n_0 + 2n)(n_0 - \gamma + 2)\right)\alpha^2}{\left(x^2 + (n_0 + 2n)^2\alpha^2\right)\left(x^2 + (n_0 + 2n + 1)^2\alpha^2\right)}.$$

It is certainly positive as desired because of $n_0 - \gamma = \beta = 1/\alpha > 0$. $\square$

Proposition 3.3, (b) and Theorem 7.2 guarantee that $B_\alpha(s,t)$ is a Schur multiplier for each $\alpha \in [-\infty, \infty]$ so that $B_\alpha(H,K)X \ (\in B(\mathcal{H}))$ makes sense for each operators H, K, X with $H, K \geq 0$, and moreover we have

$$|||B_\alpha(H,K)X||| \leq |||M_\infty(H,K)X|||. \tag{7.4}$$

7.2 Equivalence of $|||B_\alpha(H,K)X|||$ for $\alpha > 0$

In this section we investigate mutual comparison for $|||B_\alpha(H,K)X|||$ akin to Propositions 5.2 and 5.4.

Proposition 7.3. *For each $\alpha > 0$ one can find a positive constant κ_α such that*

$$|||B_\alpha(H,K)X||| \leq |||M_\infty(H,K)X||| \leq \kappa_\alpha|||B_\alpha(H,K)X|||$$

for each operators H, K, X with $H, K \geq 0$ and each unitarily invariant norm $||| \cdot |||$. In particular, the following three conditions are mutually equivalent:

(i) $|||B_\alpha(H,K)X||| < \infty$ *for some $\alpha > 0$;*
(ii) $|||M_\infty(H,K)X||| < \infty$;
(iii) $|||HX + XK||| < \infty$.

Proof. The first inequality was already pointed out (see (7.4)), and it remains to show the second. To do so, we note

$$\frac{M_\infty(e^x,1)}{B_\alpha(e^x,1)} = \frac{\max\{e^x,1\}}{\left(\frac{1+e^{\alpha x}}{2}\right)^{1/\alpha}} = 2^{1/\alpha}\frac{\max\{e^x,1\}}{(1+e^{\alpha x})^{1/\alpha}}$$

$$= 2^{1/\alpha}\left(\frac{1}{1+e^{-\alpha|x|}}\right)^{1/\alpha} = 2^{1/\alpha}\left(\frac{e^{\frac{\alpha|x|}{2}}}{e^{\frac{\alpha|x|}{2}}+e^{-\frac{\alpha|x|}{2}}}\right)^{1/\alpha}$$

$$= 2^{1/\alpha}\left(1-\frac{e^{-\frac{\alpha|x|}{2}}}{e^{\frac{\alpha|x|}{2}}+e^{-\frac{\alpha|x|}{2}}}\right)^{1/\alpha} = 2^{1/\alpha}\left(1-\frac{e^{-\frac{\alpha|x|}{2}}}{2\cosh\left(\frac{\alpha x}{2}\right)}\right)^{1/\alpha}. \quad (7.5)$$

Recalling the Taylor series expansion of $(1-z)^{1/\alpha}$ (see Lemma 7.1 and the paragraph before the lemma), we have

$$\frac{M_\infty(e^x,1)}{B_\alpha(e^x,1)} = 2^{1/\alpha}\sum_{n=0}^\infty \frac{a_n}{2^n}\times\frac{e^{-\frac{n\alpha|x|}{2}}}{\cosh^n\left(\frac{\alpha x}{2}\right)}$$

with the absolutely convergent coefficients $\sum_{n=0}^\infty 2^{-n}|a_n| < \infty$ (and $a_0 = 1$). For each $n \geq 1$ both of the functions $e^{-\frac{n\alpha|x|}{2}}$ and $1/\cosh^n\left(\frac{\alpha x}{2}\right)$ are positive definite (see (5.8), (7.3) and Example 3.6, (a)) and hence their product $e^{-\frac{n\alpha|x|}{2}}/\cosh^n\left(\frac{\alpha x}{2}\right)$ is the Fourier transform of a positive integrable function $f_n(x)$ with $\int_{-\infty}^\infty f_n(x)\,dx = 1$. In particular, by considering the sums over n's with $a_n > 0$ and $a_n < 0$ separately, we observe that $M_\infty(e^x,1)/B_\alpha(e^x,1)$ is the Fourier transform of a signed measure ν with finite total variation. Therefore, we have

$$M_\infty(H,K)X = \int_{-\infty}^\infty (Hs_H)^{ix}(B_\alpha(H,K)X)(Ks_K)^{-ix}d\nu(x),$$

and consequently the second inequality is valid with the constant $\kappa_\alpha = |\nu|(\mathbf{R})$. $\square$

From the above proof we obviously have $\kappa_\alpha = |\nu|(\mathbf{R}) \leq 2^{1/\alpha}\sum_{n=0}^\infty 2^{-n}|a_n|$. But this quantity diverges as $\alpha \searrow 0$. On the other hand, for $\alpha > 1$ one can obtain a somewhat more precise estimate. To do so, we at first point out

Lemma 7.4. *If $\alpha > 1$ and $f(x)$ is a positive definite function satisfying $0 \leq f(x) \leq 1$, then so is*

$$g(x) = 1 - (1-f(x))^{1/\alpha}.$$

Proof. As was seen in the proof of Lemma 7.1, the Taylor series expansion

$$(1-z)^{1/\alpha} = \sum_{k=0}^\infty a_n z^n$$

(with $\sum_{n=0}^\infty |a_n| < \infty$) satisfies $a_n < 0$ for each $n \geq 1$ (and $a_0 = 1$) due to $0 < 1/\alpha < 1$. By substituting $z = f(x) \in [0,1]$, we observe

$$g(x) = 1 - (1 - f(x))^{1/\alpha} = \sum_{n=1}^{\infty} (-a_n) f(x)^n.$$

The desired conclusion is clear from this expression since all the powers of $f(x)$ are positive definite.   □

For instance, with $\alpha = 2$ and $f(x) = 1/\cosh^2(x)$, we see the positive definiteness of $g(x) = 1 - |\tanh(x)|$.

Proposition 7.5. *Let* H, K *be positive operators,* $X \in B(\mathcal{H})$ *and* $||| \cdot |||$ *be any unitarily invariant norm. For* $\alpha > 1$ *we have*

$$|||B_\alpha(H,K)X||| \le |||M_\infty(H,K)X||| \le (2^{1+\frac{1}{\alpha}} - 1) |||B_\alpha(H,K)X||| \quad (7.6)$$

and

$$|||B_\alpha(H,K)X - M_\infty(H,K)X||| \le 2(2^{1/\alpha} - 1) |||B_\alpha(H,K)X|||. \quad (7.7)$$

Proof. With the special choice

$$f(x) = \frac{e^{-\frac{\alpha|x|}{2}}}{2\cosh\left(\frac{\alpha x}{2}\right)}$$

(see (7.5)) in Lemma 7.4 we observe that the function

$$g(x) = 1 - \left(\frac{1}{1 + e^{-\alpha|x|}}\right)^{1/\alpha}$$

is positive definite. Therefore, it is the Fourier transform of a positive measure with total mass $g(0) = 1 - 2^{-1/\alpha}$. We actually have

$$g(x) = 1 - 2^{-1/\alpha} \frac{M_\infty(e^x, 1)}{B_\alpha(e^x, 1)}$$

due to (7.5), and as usual we get

$$|||B_\alpha(H,K)X - 2^{-1/\alpha} M_\infty(H,K)X||| \le (1 - 2^{-1/\alpha}) |||B_\alpha(H,K)X|||. \quad (7.8)$$

From this we estimate

$$
\begin{aligned}
|||M_\infty(H,K)X||| &\le |||M_\infty(H,K)X - 2^{1/\alpha} B_\alpha(H,K)X||| \\
&\quad + 2^{1/\alpha} |||B_\alpha(H,K)X||| \\
&\le (2^{1/\alpha} - 1) |||B_\alpha(H,K)X||| + 2^{1/\alpha} |||B_\alpha(H,K)X||| \\
&= (2^{1+\frac{1}{\alpha}} - 1) |||B_\alpha(H,K)X|||,
\end{aligned}
$$

which (together with (7.4)) shows (7.6). On the other hand, (7.7) is shown from (7.8) as follows:

$$
\begin{aligned}
|||M_\infty(H,K)X &- B_\alpha(H,K)X||| \\
&\le |||M_\infty(H,K)X - 2^{1/\alpha} B_\alpha(H,K)X||| + (2^{1/\alpha} - 1) |||B_\alpha(H,K)X||| \\
&\le 2(2^{1/\alpha} - 1) |||B_\alpha(H,K)X|||.
\end{aligned}
$$

□

7.3 Norm continuity in parameter

Our goal in the section is to show the following norm continuity (and related results):

Theorem 7.6. *Let $H, K, X \in B(\mathcal{H})$ with $H, K \geq 0$, and $||| \cdot |||$ be a unitarily invariant norm. If $|||M_\infty(H, K)X||| < \infty$ (see Proposition 7.3), then one gets*

$$\lim_{\alpha \to \alpha_0} |||B_\alpha(H, K)X - B_{\alpha_0}(H, K)X||| = 0$$

for each $\alpha_0 \in [0, \infty]$.

Proposition 7.5 yields the case $\alpha_0 = \infty$ in Theorem 7.6, and hence it remains to show the case $\alpha_0 \in [0, \infty)$. The proof for the case $\alpha_0 \in (0, \infty)$ is not so hard while we will make use of a certain uniform integrability (as in the Vitali convergence theorem) to deal with the case $\alpha_0 = 0$ (see (7.11)).
   For $\alpha > 0$ the proof of Theorem 7.2 shows that

$$\hat{\psi}_\alpha(x) = \frac{B_\alpha(e^x, 1)}{M_\infty(e^x, 1)} - 2^{-1/\alpha} = 2^{-1/\alpha} \left(\left(1 + e^{-\alpha|x|}\right)^{1/\alpha} - 1 \right)$$

$$\left(= 2^{-1/\alpha} \phi_\alpha(x) \quad \text{(see (7.2))} \right)$$

with a positive integrable function $\psi_\alpha(x)$. For the limiting case $\alpha = 0$ we have

$$\hat{\psi}_0(x) = \frac{B_0(e^x, 1)}{M_\infty(e^x, 1)} = e^{-|x|/2} \left(= \lim_{\alpha \searrow 0} \hat{\psi}_\alpha(x) \right)$$

with $\psi_0(x) = \frac{1}{2\pi}\left(x^2 + \frac{1}{4}\right)^{-1}$ (see (5.8) and (7.3)). At first we compute

$$\frac{\partial}{\partial \alpha} \left(1 + e^{-\alpha|x|}\right)^{1/\alpha}$$

$$= -\left(1 + e^{-\alpha|x|}\right)^{1/\alpha} \left(\frac{|x|e^{-\alpha|x|}}{\alpha(1 + e^{-\alpha|x|})} + \frac{\log\left(1 + e^{-\alpha|x|}\right)}{\alpha^2} \right) < 0,$$

and hence

Lemma 7.7. *The function $\phi_\alpha(x)$ is monotone decreasing in $\alpha > 0$.*

The assertion (ii) of the next lemma will be proved after we prepare a few lemmas.

Lemma 7.8.

(i) $\lim_{\alpha \to \alpha_0} \|\hat{\psi}_\alpha - \hat{\psi}_{\alpha_0}\|_2 = 0$ *for $\alpha_0 > 0$,*   (ii) $\lim_{\alpha \searrow 0} \|\hat{\psi}_\alpha - \hat{\psi}_0\|_2 = 0.$

We at first assume $\alpha_0 > 0$ and choose a positive integer n_0 with $\frac{1}{n_0} < \alpha_0$. For $\alpha \geq \frac{1}{n_0}$, thanks to Lemma 7.7 we observe

$$\hat{\psi}_\alpha(x) = 2^{-1/\alpha}\phi_\alpha(x) \leq \phi_\alpha(x) \leq \phi_{\frac{1}{n_0}}(x).$$

Notice

$$\phi_{\frac{1}{n_0}}(x) = \left(1 + e^{-\frac{|x|}{n_0}}\right)^{n_0} - 1 = \sum_{k=1}^{n_0} \binom{n_0}{k} e^{-\frac{k}{n_0}|x|}$$

$$\leq (2^{n_0} - 1)e^{-\frac{|x|}{n_0}} \in L^2(\mathbf{R}).$$

When $\alpha \to \alpha_0$ (with $\alpha \geq \frac{1}{n_0}$), we obviously have $\hat{\psi}_\alpha(x) \to \hat{\psi}_{\alpha_0}(x)$ for each $x \in \mathbf{R}$ so that (i) in Lemma 7.8 follows from the Lebesgue dominated convergence theorem.

We next deal with the case $\alpha_0 = 0$ (i.e., Lemma 7.8, (ii)). We begin with the special case

$$\lim_{n\to\infty} \|\hat{\psi}_{\frac{1}{n}} - \hat{\psi}_0\|_2 = 0.$$

Since all the relevant functions here are even, what we really have to show is

$$\lim_{n\to\infty} \int_0^\infty |f_n(t) - f_\infty(t)|^2 dt = 0 \tag{7.9}$$

where

$$f_n(t) = 2^{-n}\left(\left(1 + e^{-\frac{t}{n}}\right)^n - 1\right) \quad \text{and} \quad f_\infty(t) = e^{-t/2} \quad (t \geq 0).$$

To show (7.9) we use the binomial expansion

$$f_n(t) = 2^{-n}\sum_{k=1}^n \binom{n}{k} e^{-\frac{k}{n}t}.$$

Choose and fix $\delta > 0$ small, and we split the sum $\sum_{k=1}^n$ into the following two parts:

$$\begin{cases} \sum' : \text{summation over } k \in \{1, 2, \ldots, n\} \text{ with } \frac{k}{n} > \delta, \\ \sum'' : \text{summation over } k \in \{1, 2, \ldots, n\} \text{ with } \frac{k}{n} \leq \delta. \end{cases}$$

We observe

$$f_n(t) \leq 2^{-n}\sum{}' \binom{n}{k} e^{-\delta t} + 2^{-n}\sum{}'' \binom{n}{k} e^{-\frac{k}{n}t}$$

$$\leq e^{-\delta t} + 2^{-n}\sum{}'' \binom{n}{k} e^{-\frac{k}{n}t}.$$

By making use of the obvious fact $0 \leq f_n(t) \leq 1$, from the above inequality we estimate

$$\int_M^\infty f_n(t)^2 dt \leq \int_M^\infty f_n(t)\, dt \leq \int_M^\infty e^{-\delta t} dt + 2^{-n} \sum{}'' \binom{n}{k} \int_M^\infty e^{-\frac{k}{n}t} dt$$

$$= \frac{1}{\delta} e^{-\delta M} + 2^{-n} \sum{}'' \binom{n}{k} \frac{n}{k} e^{-\frac{k}{n}M}$$

$$\leq \frac{1}{\delta} e^{-\delta M} + 2^{-n} \sum{}'' \binom{n}{k} \frac{n}{k} \tag{7.10}$$

for $M > 0$ (to be specified shortly). Note that the second factor (containing binomial coefficients) in the above far right side is no longer depending upon M.

Lemma 7.9. *When $\delta > 0$ is small enough, we have*

$$\lim_{n\to\infty} 2^{-n} \sum{}'' \binom{n}{k} \frac{n}{k} = 0.$$

Proof. Based on the Stirling formula we estimate

$$\log\left(2^{-n} \binom{n}{k} \frac{n}{k}\right)$$
$$= -n\log 2 + \log(n!) - \log(k!) - \log((n-k)!) + \log n - \log k$$
$$= -n\log 2 + n\log n - n + \frac{1}{2}\log n$$
$$\quad - k\log k + k - \frac{1}{2}\log k - (n-k)\log(n-k) + (n-k) - \frac{1}{2}\log(n-k)$$
$$\quad + \log n - \log k + O(1)$$
$$\leq -n\log 2 - k\log\left(\frac{k}{n}\right) - (n-k)\log\left(\frac{n-k}{n}\right) + \frac{3}{2}\log n + O(1)$$

$$= -n\left(\log 2 + \frac{k}{n}\log\left(\frac{k}{n}\right) + \left(1 - \frac{k}{n}\right)\log\left(1 - \frac{k}{n}\right)\right) + \frac{3}{2}\log n + O(1).$$

We set $\theta(x) = x\log x$, and notice $\theta(0) = \theta(1) = 0$ and $\theta(x) < 0$ for $x \in (0,1)$. Choose and fix $0 < \varepsilon_0 < \log 2$, and assume that $\delta > 0$ is chosen small enough in such a way that

$$\theta(x) + \theta(1-x) \geq -\varepsilon_0 \quad \text{for } 0 < x \leq \delta$$

is guaranteed. Then, since $\frac{k}{n} \leq \delta$ for k's appearing in the sum $\sum''$, we get

$$\log 2 + \frac{k}{n}\log\left(\frac{k}{n}\right) + \left(1 - \frac{k}{n}\right)\log\left(1 - \frac{k}{n}\right) \geq \log 2 - \varepsilon_0.$$

Thus, from the preceding estimate we get

$$\log\left(2^{-n} \binom{n}{k} \frac{n}{k}\right) \leq -n(\log 2 - \varepsilon_0) + \frac{3}{2}\log n + O(1),$$

that is,

$$2^{-n} \binom{n}{k} \frac{n}{k} \leq K \times n^{3/2} \left(\frac{e^{\varepsilon_0}}{2} \right)^n \quad \text{(as long as } \frac{k}{n} \leq \delta \text{)}$$

for some constant K. Therefore, we conclude

$$\frac{1}{2^n} \sum {}'' \binom{n}{k} \frac{n}{k} \leq K \times n^{5/2} \left(\frac{e^{\varepsilon_0}}{2} \right)^n,$$

showing the desired convergence due to $\frac{e^{\varepsilon_0}}{2} < 1$. □

Lemma 7.10. *The L^2-convergence (7.9) is valid, that is, we have*

$$\lim_{n \to \infty} \| \hat{\psi}_{\frac{1}{n}} - \hat{\psi}_0 \|_2 = 0.$$

Proof. Let $\varepsilon > 0$. We claim the following uniform integrability: one can find an integer N and a positive M such that

$$\int_M^\infty f_n(t)^2 dt < \varepsilon \quad \text{for each } n \geq N. \tag{7.11}$$

In fact, we recall the estimate (7.10), and fix a small $\delta > 0$ so that Lemma 7.9 is valid. From this lemma, we have $2^{-n} \sum '' \binom{n}{k} \frac{n}{k} < \varepsilon/2$ for n large enough. Then, one can choose $M > 0$ large enough so that $\frac{1}{\delta} e^{-\delta M} < \varepsilon/2$.

We estimate

$$\int_0^\infty |f_n(t) - f_\infty(t)|^2 dt$$

$$\leq \int_0^M |f_n(t) - f_\infty(t)|^2 dt + \int_M^\infty (f_n(t) + f_\infty(t))^2 dt$$

$$\leq \int_0^M |f_n(t) - f_\infty(t)|^2 dt + 2 \int_M^\infty f_n(t)^2 dt + 2 \int_M^\infty f_\infty(t)^2 dt.$$

Fatou's lemma shows

$$\int_M^\infty f_\infty(t)^2 dt \leq \liminf_{n \to \infty} \int_M^\infty f_n(t)^2 dt.$$

Thus, from the estimates so far and (7.11) we get

$$\int_0^\infty |f_n(t) - f_\infty(t)|^2 dt \leq \int_0^M |f_n(t) - f_\infty(t)|^2 dt + 4\varepsilon$$

for n large enough, and hence

$$\limsup_{n \to \infty} \int_0^\infty |f_n(t) - f_\infty(t)|^2 dt \leq \limsup_{n \to \infty} \int_0^M |f_n(t) - f_\infty(t)|^2 dt + 4\varepsilon.$$

Note that lim sup in the right-hand side is 0 since $f_n(t) \leq 1$ enables us to use the Lebesgue dominated convergence theorem on the finite interval $[0, M]$. Since $\varepsilon > 0$ is arbitrary, we are done. □

Proof of Lemma 7.8, (ii). What we have to show is $\lim_{k\to\infty} \|\hat{\psi}_{\alpha_k} - \hat{\psi}_0\|_2 = 0$ for each decreasing sequence $\{\alpha_k\}_{k=1,2,\cdots}$ converging to 0. For each k one takes the natural number n_k such that $n_k - 1 < 1/\alpha_k \leq n_k$ so that $\{n_k\}_{k=1,2,\cdots}$ is an increasing sequence tending to ∞. Notice $2^{-n_k} \leq 2^{-1/\alpha_k} < 2 \times 2^{-n_k}$. Hence, from Lemma 7.7 we have

$$\hat{\psi}_{\alpha_k}(x) \leq 2 \times 2^{-n_k}\left(\left(1 + e^{-\alpha_k|x|}\right)^{1/\alpha_k} - 1\right)$$

$$\leq 2 \times 2^{-n_k}\left(\left(1 + e^{-\frac{|x|}{n_k}}\right)^{n_k} - 1\right) = 2 \times \hat{\psi}_{\frac{1}{n_k}}(x).$$

Therefore, the L^p-version of the extended Lebesgue convergence theorem (see [25, p. 122] or [26, Theorem 3.6]) and Lemma 7.10 yield $\lim_{k\to\infty} \|\hat{\psi}_{\alpha_k} - \hat{\psi}_0\|_2 = 0$ as desired. $\square$

Proof of Theorem 7.6. As was pointed out right after the theorem, we may and do assume $\alpha_0 \in [0, \infty)$. Lemmas 5.8 and 7.8 yield

$$\lim_{\alpha\to\alpha_0} \|\psi_\alpha - \psi_{\alpha_0}\|_1 = 0 \quad (\alpha_0 > 0) \quad \text{and} \quad \lim_{\alpha\searrow 0} \|\psi_\alpha - \psi_0\|_1 = 0.$$

Thus, from the integral expressions

$$B_\alpha(H, K)X = 2^{-1/\alpha} M_\infty(H, K)X$$
$$+ \int_{-\infty}^{\infty} (Hs_H)^{ix}(M_\infty(H, K)X)(Ks_K)^{-ix}\psi_\alpha(x)\, dx,$$

$$B_0(H, K)X = \int_{-\infty}^{\infty} (Hs_H)^{ix}(M_\infty(H, K)X)(Ks_K)^{-ix}\psi_0(x)\, dx,$$

we get

$$\lim_{\alpha\to\alpha_0} |||B_\alpha(H, K)X - B_{\alpha_0}(H, K)X||| = 0$$

and

$$\lim_{\alpha\searrow 0} |||B_\alpha(H, K)X - B_0(H, K)X||| = 0.$$

It remains to show

$$\lim_{\alpha\nearrow 0} |||B_\alpha(H, K)X - B_0(H, K)X||| = 0.$$

Thus, we assume $\alpha < 0$ and set

$$\varphi_\alpha(x) = \frac{B_\alpha(e^x, 1)}{M_\infty(e^x, 1)} = \left(\frac{1 + e^{-\alpha|x|}}{2}\right)^{1/\alpha} \quad \text{(see (7.1))}$$

$$= e^{-|x|/2} \cosh^{1/\alpha}\left(\frac{\alpha|x|}{2}\right) = e^{-|x|/2}\left(\frac{1}{\cosh\left(\frac{(-\alpha)x}{2}\right)}\right)^{\frac{1}{-\alpha}},$$

$$\varphi_0(x) = \frac{B_0(e^x, 1)}{M_\infty(e^x, 1)} = e^{-|x|/2}.$$

Then, φ_α is a positive definite function and $\varphi_\alpha = \hat{\psi}_\alpha$ ($\alpha \leq 0$) with a positive integrable function ψ_α (see (5.8) and the proof of [39, Proposition 3.3], and also see §A.6). Since $\varphi_\alpha(x) \leq \varphi_0(x) = e^{-|x|/2} \in L^2(\mathbf{R})$ and $\lim_{\alpha \nearrow 0} \varphi_\alpha(x) = \varphi_0(x)$, we have $\lim_{\alpha \nearrow 0} \|\phi_\alpha - \phi_0\|_2 = 0$ by the Lebesgue dominated convergence theorem. Since

$$\int_{-\infty}^{\infty} \psi_\alpha(x)\,dx = \varphi_\alpha(0) = 1 \qquad (\alpha \leq 0),$$

Lemma 5.8 shows

$$\lim_{\alpha \nearrow 0} \|\psi_\alpha - \psi_0\|_1 = 0$$

and the desired convergence follows from the integral expression

$$B_\alpha(H, K)X = \int_{-\infty}^{\infty} (H s_H)^{ix} (M_\infty(H, K)X)(K s_K)^{-ix} \psi_\alpha(x)\,dx \qquad (\alpha \leq 0).$$

$\square$

Recall $B_0(H, K)X = G(H, K)X = H^{1/2}XK^{1/2}$, the geometric mean. For operator means $B_\alpha(H, K)X$ with $\alpha < 0$ we have

Proposition 7.11. If $\||H^{1/2}XK^{1/2}\|| < \infty$, then

$$\lim_{\alpha \to \alpha_0} \||B_\alpha(H, K)X - B_{\alpha_0}(H, K)X\|| = 0$$

for every $\alpha_0 \in [-\infty, 0)$.

Proof. We set

$$\varphi_\alpha(x) = \frac{B_\alpha(e^x, 1)}{G(e^x, 1)} = e^{-x/2}\left(\frac{e^{\alpha x} + 1}{2}\right)^{1/\alpha}$$

$$= \cosh^{1/\alpha}\left(\frac{\alpha x}{2}\right) = \left(\frac{1}{\cosh\left(\frac{(-\alpha)x}{2}\right)}\right)^{\frac{1}{-\alpha}}$$

for $\alpha < 0$. Then, φ_α is a positive definite function and $\varphi_\alpha = \hat{\psi}_\alpha$ with a positive function $\psi_\alpha \in L^1(\mathbf{R})$ (see the proof of [39, Proposition 3.3] or §A.6). Note that $\varphi_\alpha(x)$ is monotone increasing in α because so is $B_\alpha(e^x, 1)$ as noted just after (7.4).

Let us assume $\alpha_0 \neq -\infty$. When $|\alpha - \alpha_0| \leq \frac{-\alpha_0}{2}$, we have $\alpha \leq \frac{\alpha_0}{2}$ and consequently $0 \leq \varphi_\alpha(x) \leq \varphi_{\frac{\alpha_0}{2}}(x)$ with $\varphi_{\frac{\alpha_0}{2}} \in L^1(\mathbf{R})$. Of course we have $\lim_{\alpha \to \alpha_0} \varphi_\alpha(x) = \varphi_{\alpha_0}(x)$, and the Lebesgue dominated convergence theorem implies

$$\lim_{\alpha \to \alpha_0} \|\varphi_\alpha - \varphi_{\alpha_0}\|_1 = 0.$$

Hence, for each sequence $\{\alpha_k\}_{k=1,2,\cdots}$ converging to α_0 we get

$$|\psi_{\alpha_k}(x) - \psi_{\alpha_0}(x)|$$

$$\leq \frac{1}{2\pi}\left|\int_{-\infty}^{\infty}(\varphi_{\alpha_k}(y) - \varphi_{\alpha_0}(y))e^{ixy}\,dy\right| \leq \frac{1}{2\pi} \times \|\varphi_{\alpha_k} - \varphi_{\alpha_0}\|_1 \longrightarrow 0$$

for a.e. $x \in \mathbf{R}$. Since

$$\int_{-\infty}^{\infty} \psi_\alpha(x)\,dx = \varphi_\alpha(0) = 1 \quad \text{for each } \alpha < 0,$$

we have

$$\lim_{\alpha \to \alpha_0} \|\psi_\alpha - \psi_{\alpha_0}\|_1 = 0$$

as usual, and the required convergence can be seen from the integral expression

$$B_\alpha(H, K)X = \int_{-\infty}^{\infty} (H s_H)^{ix}(H^{1/2}XK^{1/2})(K s_K)^{-ix}\psi_\alpha(x)\,dx.$$

Obviously the same proof works for $\alpha_0 = -\infty$ as well with $\varphi_{-\infty}(x) = e^{-|x|/2}$ and $\psi_{-\infty}(x) = \frac{1}{2\pi}(x^2 + \frac{1}{4})^{-1}$ (see (5.8) and (7.3)), and details are left to the reader. □

Alternative proof of Proposition 7.11. As in the above proof we set

$$\varphi_\alpha(x) = \left(\frac{1}{\cosh\left(\frac{(-\alpha)x}{2}\right)}\right)^{\frac{1}{-\alpha}} \quad (\alpha < 0) \quad \text{and} \quad \varphi_{-\infty}(x) = e^{-|x|/2}.$$

Then, φ_α is a positive definite function. In fact, we have $\varphi_\alpha = \hat\psi_\alpha$ with the function

$$\psi_\alpha(x) = \frac{1}{2\pi} \times \frac{2^{\frac{1}{-\alpha}}}{(-\alpha)\Gamma\left(\frac{1}{-\alpha}\right)} \times \left|\Gamma\left(\frac{1}{2(-\alpha)} + \frac{ix}{-\alpha}\right)\right|^2 \geq 0.$$

Details are worked out in §A.6 (see (A.10)), and this explicit form was pointed out in [13].

The obvious continuity of the Γ-function $\Gamma(z)$ shows

$$\lim_{\alpha \to \alpha_0} \psi_\alpha(x) = \psi_{\alpha_0}(x) \tag{7.12}$$

for $\alpha_0 < 0$. This fact remains valid for $\alpha_0 = -\infty$ as well. In fact, thanks to $\Gamma(z+1) = z\Gamma(z)$ we compute

$$\lim_{\alpha \to -\infty} \psi_\alpha(x) = \frac{1}{2\pi}\lim_{\beta \searrow 0}\left(\frac{2^\beta \beta}{\Gamma(\beta)} \times \left|\Gamma\left(\frac{\beta}{2} + ix\beta\right)\right|^2\right)$$

$$= \frac{1}{2\pi}\lim_{\beta \searrow 0}\left(\frac{2^\beta \beta^2}{\Gamma(\beta+1)} \times \left|\Gamma\left(\frac{\beta}{2} + ix\beta\right)\right|^2\right)$$

$$= \frac{1}{2\pi}\lim_{\beta \searrow 0}\beta^2\left|\Gamma\left(\frac{\beta}{2} + ix\beta\right)\right|^2.$$

Recall that $0, -1, -2, -3, \ldots$ are simple poles of $\Gamma(z)$ (see [80] for example). The residue at 0 is 1 so that near the origin $\Gamma(z)$ is of the form $\frac{1}{z} + f(z)$ with a holomorphic function $f(z)$. This means $\Gamma(\frac{\beta}{2} + ix\beta) \sim (\frac{\beta}{2} + ix\beta)^{-1}$ for β small, and we conclude

$$\lim_{\alpha \to -\infty} \psi_\alpha(x) = \frac{1}{2\pi} \lim_{\beta \searrow 0} \frac{\beta^2}{|\frac{\beta}{2} + ix\beta|^2} = \frac{1}{2\pi} \times \frac{1}{|\frac{1}{2} + ix|^2} = \frac{1}{2\pi} \times \frac{1}{x^2 + \frac{1}{4}}.$$

This limit function is exactly $\psi_{-\infty}(x)$ (see (5.8) and (7.3)), and hence (7.12) has been checked for $\alpha_0 = -\infty$.

We have

$$\int_{-\infty}^{\infty} \psi_\alpha(x)\, dx = \varphi_\alpha(0) = 1 \quad \text{for each } \alpha \in [-\infty, 0).$$

Hence, (7.12) and the extended Lebesgue dominated convergence theorem yield

$$\lim_{\alpha \to \alpha_0} \|\psi_\alpha - \psi_{\alpha_0}\|_1 = 0,$$

and as usual the required convergence can be seen from the integral expression

$$B_\alpha(H, K)X = \int_{-\infty}^{\infty} (Hs_H)^{ix} (H^{1/2} X K^{1/2})(Ks_K)^{-ix} \psi_\alpha(x)\, dx.$$

$\square$

From Theorem 7.6 and Proposition 7.11 we get

Corollary 7.12. *For each $H, K, X \in B(\mathcal{H})$ with $H, K \geq 0$ and each $-\infty \leq \alpha_0 \leq \infty$ we always have*

$$\lim_{\alpha \to \alpha_0} \|B_\alpha(H, K)X - B_{\alpha_0}(H, K)X\| = 0$$

in the operator norm $\|\cdot\|$.

7.4 Notes and references

The binomial means $\{B_\alpha\}_{-\infty \leq \alpha \leq \infty}$ and the corresponding means $B_\alpha(H, K)X$ were studied in [39]. Only matrices were dealt with there so that the majorization $B_\alpha \preceq M_\infty$ (Theorem 7.2) was not necessary. This majorization (together with Proposition 3.3, (b)) makes the notion of operator means $B_\alpha(H, K)X$ legitimate for Hilbert space operators.

We note that $B_\alpha(s, t)$ is monotone increasing in $\alpha \in [-\infty, \infty]$ for each fixed $s, t > 0$. Indeed, when $0 < \alpha < \beta$, the concavity of the function $t^{\alpha/\beta}$ $(t > 0)$ gives

$$\frac{s^\alpha + t^\alpha}{2} \leq \left(\frac{s^\beta + t^\beta}{2} \right)^{\alpha/\beta},$$

i.e., $B_\alpha(s,t) \le B_\beta(s,t)$. The case $\alpha < \beta < 0$ is similarly checked. However, the comparison $B_\alpha \preceq B_\beta$ (similar to (5.2) and (6.2)) for general $-\infty \le \alpha < \beta \le \infty$ is an interesting open problem. The following partial results were obtained in [39, Proposition 3.3]:

(i)   for $\alpha \ge 0$ we have $B_0 \ (= M_{1/2}) \preceq B_\alpha \ (\preceq B_\infty \ (= M_\infty))$;

(ii) we have $B_{\frac{1}{m}} \preceq B_{\frac{1}{n}}$ as long as $n \ (\in \mathbf{N})$ divides m.

Certain alternating sums of operators

In this chapter we will deal with alternating sums

$$H^{\frac{1}{2}}XK^{\frac{1}{2}}$$

$$H^{\frac{1}{3}}XK^{\frac{2}{3}} - H^{\frac{2}{3}}XK^{\frac{1}{3}}$$

$$H^{\frac{1}{4}}XK^{\frac{3}{4}} - H^{\frac{2}{4}}XK^{\frac{2}{4}} + H^{\frac{3}{4}}XK^{\frac{1}{4}}$$

$$H^{\frac{1}{5}}XK^{\frac{4}{5}} - H^{\frac{2}{5}}XK^{\frac{3}{5}} + H^{\frac{3}{5}}XK^{\frac{2}{5}} - H^{\frac{4}{5}}XK^{\frac{1}{5}}$$

$$\cdots$$

$$XK - HX$$

$$XK - H^{\frac{1}{2}}XK^{\frac{1}{2}} + HX$$

$$XK - H^{\frac{1}{3}}XK^{\frac{2}{3}} + H^{\frac{2}{3}}XK^{\frac{1}{3}} - HX$$

$$XK - H^{\frac{1}{4}}XK^{\frac{3}{4}} + H^{\frac{2}{4}}XK^{\frac{2}{4}} - H^{\frac{3}{4}}XK^{\frac{1}{4}} + HX$$

$$\cdots,$$

and investigate behavior of unitarily invariant norms of these operators such as mutual comparison, uniform bounds (independent of n, m), monotonicity and so on (in §8.2 and §8.3). For convenience we set

$$\mathbf{A}(n) = \sum_{k=1}^{n} (-1)^{k-1} H^{\frac{k}{n+1}} XK^{\frac{n+1-k}{n+1}} \quad (n = 1, 2, 3, \cdots),$$

$$\mathbf{B}(m) = \sum_{k=0}^{m-1} (-1)^{k} H^{\frac{k}{m-1}} XK^{\frac{m-1-k}{m-1}} \quad (m = 2, 3, 4, \cdots),$$

and these notations will be kept throughout. We note

$$\mathbf{B}(m) = \begin{cases} HX + XK - \mathbf{A}(m-2) & \text{for } m = 3, 5, 7, \cdots, \\ -HX + XK - \mathbf{A}(m-2) & \text{for } m = 4, 6, 8, \cdots. \end{cases} \quad (8.1)$$

The nature of the above two series of operators depends strongly on parities of n and m, and it is quite obvious that we will have to treat odd and even cases separately.

8.1 Preliminaries

For $n = 1, 2, \cdots$ and $m = 2, 3, \cdots$ we set

$$a_n(s,t) = \sum_{k=1}^{n} (-1)^{k-1} s^{\frac{k}{n+1}} t^{\frac{n+1-k}{n+1}} \quad \text{and} \quad b_m(s,t) = \sum_{k=0}^{m-1} (-1)^k s^{\frac{k}{m-1}} t^{\frac{m-1-k}{m-1}}$$

$(s, t \geq 0)$ as scalar "means" corresponding to $\mathbf{A}(n)$ and $\mathbf{B}(m)$. For $s, t > 0$ we compute

$$a_n(s,t) = \frac{s^{\frac{1}{n+1}} t^{\frac{n}{n+1}} \left(1 - (-1)^n \left(\frac{s}{t}\right)^{\frac{n}{n+1}}\right)}{1 + \left(\frac{s}{t}\right)^{\frac{1}{n+1}}} = \frac{t \left(\frac{s}{t}\right)^{\frac{1}{n+1}} \left(1 - (-1)^n \left(\frac{s}{t}\right)^{\frac{n}{n+1}}\right)}{1 + \left(\frac{s}{t}\right)^{\frac{1}{n+1}}}$$

$$= t \left(\frac{s}{t}\right)^{\frac{1}{2}} \times \frac{\left(\frac{s}{t}\right)^{-\frac{1}{2} \times \frac{n}{n+1}} - (-1)^n \left(\frac{s}{t}\right)^{\frac{1}{2} \times \frac{n}{n+1}}}{\left(\frac{s}{t}\right)^{-\frac{1}{2} \times \frac{1}{n+1}} + \left(\frac{s}{t}\right)^{\frac{1}{2} \times \frac{1}{n+1}}}$$

$$= (st)^{\frac{1}{2}} \times \frac{\left(\frac{s}{t}\right)^{-\frac{1}{2} \times \frac{n}{n+1}} - (-1)^n \left(\frac{s}{t}\right)^{\frac{1}{2} \times \frac{n}{n+1}}}{\left(\frac{s}{t}\right)^{-\frac{1}{2} \times \frac{1}{n+1}} + \left(\frac{s}{t}\right)^{\frac{1}{2} \times \frac{1}{n+1}}}.$$

Note that the denominator can be always expressed in terms of the hyperbolic cosine function while for the numerator the hyperbolic sine function is also needed for n even. Exactly the same computations yield

$$b_m(s,t) = (st)^{\frac{1}{2}} \times \frac{\left(\frac{s}{t}\right)^{-\frac{1}{2} \times \frac{m}{m-1}} - (-1)^m \left(\frac{s}{t}\right)^{\frac{1}{2} \times \frac{m}{m-1}}}{\left(\frac{s}{t}\right)^{-\frac{1}{2} \times \frac{1}{m-1}} + \left(\frac{s}{t}\right)^{\frac{1}{2} \times \frac{1}{m-1}}}.$$

These formulas will be freely and repeatedly used. We note the homogeneity

$$a_n(rs, rt) = r a_n(s,t), \quad b_m(rs, rt) = r b_m(s,t)$$

(with $r \geq 0$) and

$$a_n(t, s) = (-1)^{n+1} a_n(s,t), \quad b_m(t, s) = (-1)^{m+1} b_m(s,t)$$

(see Proposition 8.2, (iii)).

We will repeatedly make use of the positive definiteness of the following functions (see §6.3, **1**):

$$\frac{1}{\cosh(\alpha x)}, \quad \frac{\cosh(\beta x)}{\cosh(\alpha x)}, \quad \frac{\sinh(\beta x)}{\sinh(\alpha x)}$$

with $0 < \beta < \alpha$ (as was done in preceding chapters). The next observation is also useful.

Lemma 8.1. *For* $\alpha, \beta > 0$ *one can find a signed measure* ν *on* $\mathbf{R}$ *such that*

$$\frac{\cosh((\alpha + \beta)x)}{\cosh(\alpha x)\cosh(\beta x)} = \hat{\nu}(x) \left(= \int_{-\infty}^{\infty} e^{ixy} d\nu(y) \right)$$

with $|\nu|(\mathbf{R}) \le 5$.

Proof. By the addition rule for the hyperbolic cosine function we observe

$$\frac{\cosh((\alpha + \beta)x)}{\cosh(\alpha x)\cosh(\beta x)} = 1 + \frac{\sinh(\alpha x)\sinh(\beta x)}{\cosh(\alpha x)\cosh(\beta x)}$$
$$= 1 + |\tanh(\alpha x)| \times |\tanh(\beta x)|.$$

We set

$$p(x) = 1 - |\tanh(\alpha x)|, \quad q(x) = 1 - |\tanh(\beta x)|.$$

As was shown in Lemma 7.4 (see the paragraph right after the lemma), they are positive definite and we observe

$$\frac{\cosh((\alpha + \beta)x)}{\cosh(\alpha x)\cosh(\beta x)} = 1 + (1 - p(x))(1 - q(x))$$
$$= (2 + p(x)q(x)) - (p(x) + q(x)).$$

Note that both of $2 + p(x)q(x)$ and $p(x) + q(x)$ are positive definite with

$$2 + p(0)q(0) = 3 \quad \text{and} \quad p(0) + q(0) = 2.$$

By the Bochner theorem there exist positive measures ν_1, ν_2 with the Fourier transforms $2 + p(x)q(x)$, $p(x) + q(x)$ respectively and $\nu_1(\mathbf{R}) = 3$, $\nu_2(\mathbf{R}) = 2$. Hence, the difference measure $\nu = \nu_1 - \nu_2$ does the job. $\square$

Let $M(s,t), N(s,t)$ be continuous functions on $[0, \infty) \times [0, \infty)$ satisfying the homogeneity condition

$$M(rs, rt) = rM(s,t) \quad \text{and} \quad N(rs, rt) = rN(s,t) \quad \text{(for } r \ge 0),$$

from which we obviously have

$$\begin{cases} M(s,0) = sM(1,0), \ N(s,0) = sN(1,0), \\ M(0,t) = tM(0,1), \ N(0,t) = tN(0,1). \end{cases} \tag{8.2}$$

We further assume that $M(s,t), N(s,t)$ are Schur multipliers (relative to any pair (H, K)) in the sense explained in §2.1. This assumption is not harmful at all because the proposition below will be applied for a_n's and b_m's (which are obviously Schur multipliers). The corresponding "operator means" will be denoted by $M(H, K)X, N(H, K)X$ as in Chapter 3.

Many integral expressions were obtained in Chapter 3 to establish norm inequalities. In particular, Theorem 3.4 deals with the symmetric homogeneous means; the proof is still valid if $M(s,t), N(s,t)$ are symmetric homogeneous functions and ν is a signed measure such that $M(e^x,1)/N(e^x,1) = \hat{\nu}(x)$. The next proposition (as well as its proof) is a variant of this result in the non-symmetric case. The part (iii) plays a fundamental role in the present chapter while the part (ii) will be used in our forthcoming article [55].

Proposition 8.2. *We assume that homogeneous Schur multipliers as above satisfy*

$$\frac{M(e^x,1)}{N(e^x,1)} = \hat{\nu}(x)$$

with a signed measure ν on $\mathbf{R}$.

(i)   *When $H, K \geq 0$ are non-singular, we have*

$$M(H,K)X = \int_{-\infty}^{\infty} H^{ix}(N(H,K)X)K^{-ix}d\nu(x).$$

(ii)   *When $M(1,0) = M(0,1) = 0$, we have*

$$M(H,K)X = \int_{-\infty}^{\infty} (Hs_H)^{ix}(N(H,K)X)(Ks_K)^{-ix}d\nu(x).$$

(iii)   *When $M(s,t) = -M(t,s)$ and $N(s,t) = -N(t,s)$, we have*

$$M(H,K)X = \int_{\{x \neq 0\}} (Hs_H)^{ix}(N(H,K)X)(Ks_K)^{-ix}d\nu(x)$$
$$+\nu(\{0\})N(H,K)X.$$

Proof. The assertions (i), (ii) directly follow from Proposition 2.11, and it remains to prove (iii). To do so, we firstly note

$$\frac{M(s,t)}{N(s,t)} = \frac{M(t,s)}{N(t,s)} \tag{8.3}$$

by the assumption in (iii). Secondly, Lemma 2.9 and (8.2) show

$$N(H,K)X = s_H(N(H,K)X)s_K$$
$$+N(1,0)HX(1-s_K) + N(0,1)(1-s_H)XK \tag{8.4}$$

(which is a replacement of (3.10) in the proof of Theorem 3.4).
    We claim

$$M(1,0) = \nu(\{0\})N(1,0) \quad \text{and} \quad M(0,1) = \nu(\{0\})N(0,1). \tag{8.5}$$

To see the claim, we begin by noting

$$M(s,t) = tM(s/t,1) = tN(s/t,1)\int_{-\infty}^{\infty} e^{ix(\log s - \log t)}d\nu(x)$$

$$= N(s,t)\int_{-\infty}^{\infty}(s/t)^{ix}d\nu(x) \quad \text{(for } s,t > 0)$$

(thanks to the assumption and the homogeneity). Thus, by the obvious continuity we get

$$|M(1,0)| \le |\nu|(\mathbf{R}) \times |N(1,0)|. \tag{8.6}$$

Firstly, if $N(1,0) = 0$ (or equivalently $N(0,1) = 0$), then we get $M(1,0) = M(0,1) = 0$ by (8.6) and (8.5) is certainly valid. Secondly, let us assume $N(1,0) \ne 0$. Since ν is a symmetric measure (by (8.3)), we have

$$\lim_{x\to\pm\infty}\hat\nu(x) = \lim_{x\to\infty}\frac{M(e^x,1)}{N(e^x,1)} = \lim_{x\to\infty}\frac{M(1,e^{-x})}{N(1,e^{-x})} = \frac{M(1,0)}{N(1,0)}.$$

Therefore, the claim (i.e., (8.5)) follows from Corollary A.8 thanks to

$$\lim_{x\to\pm\infty}\hat\nu(x) = \frac{M(1,0)}{N(1,0)} = \frac{M(0,1)}{N(0,1)}.$$

From Proposition 2.11 together with (8.5) we see

$$M(H,K)X = \int_{-\infty}^{\infty}(Hs_H)^{ix}(N(H,K)X)(Ks_K)^{-ix}d\nu(x)$$

$$+\nu(\{0\})\Big(s_H(N(H,K)X)(1-s_K) + (1-s_H)(N(H,K)X)s_K\Big).$$

Therefore, with (8.4) we compute

$$M(H,K)X$$

$$= \int_{-\infty}^{\infty}(Hs_H)^{ix}(N(H,K)X)(Ks_K)^{-ix}d\nu(x)$$

$$+\nu(\{0\})\Big(N(1,0)HX(1-s_K) + N(0,1)(1-s_H)XK\Big)$$

$$= \int_{\{x\ne 0\}}(Hs_H)^{ix}(N(H,K)X)(Ks_K)^{-ix}d\nu(x)$$

$$+\nu(\{0\})\Big(s_H(N(H,K)X)s_K$$

$$+N(1,0)HX(1-s_K) + N(0,1)(1-s_H)XK\Big)$$

$$= \int_{\{x\ne 0\}}(Hs_H)^{ix}(N(H,K)X)(Ks_K)^{-ix}d\nu(x) + \nu(\{0\})N(H,K)X,$$

showing (iii). □

The integral expressions in the proposition yield

$$|||M(H,K)X||| \le |\nu|(\mathbf{R}) \times |||N(H,K)X|||$$

for each unitarily invariant norm $|||\cdot|||$ (when one of the conditions (i), (ii), (iii) is satisfied). In fact, it follows from the Hahn decomposition $\nu = \nu_+ - \nu_-$ and Theorem A.5. In the next two sections we will deal with signed measures ν satisfying $M(e^{2x},1)/N(e^{2x},1) = \hat{\nu}(x)$ instead. This means that our integral expression is actually of the form

$$M(H,K)X = \int_{\{x \ne 0\}} (H s_H)^{\frac{ix}{2}} (N(H,K)X)(K s_K)^{-\frac{ix}{2}} d\nu(x)$$

$$+ \nu(\{0\})N(H,K)X$$

(for example in case (iii)), and hence we have the same estimate as above.

8.2 Uniform bounds for norms

In this section we investigate uniform (upper and lower) bounds for $|||\mathbf{A}(n)|||$'s and $|||\mathbf{B}(m)|||$'s. We begin with comparison between $|||\mathbf{A}(n)|||$ and $|||\mathbf{B}(m)|||$. As was remarked before, odd and even cases have to be studied separately.

Theorem 8.3.

(i)   *For $n = 1, 3, 5, \cdots$ and $m = 3, 5, 7, \cdots$ we have*

$$|||\mathbf{A}(n)||| \le |||\mathbf{B}(m)|||.$$

(ii) *For $n = 2, 4, 6, \cdots$ and $m = 2, 4, 6, \cdots$ we have*

$$\frac{n+1}{n} \times |||\mathbf{A}(n)||| \le \frac{m-1}{m} \times |||\mathbf{B}(m)|||.$$

Proof. We set

$$\alpha = \frac{1}{n+1} \quad \text{and} \quad \beta = \frac{1}{m-1}.$$

(i) We compute

$$\frac{a_n(e^{2x},1)}{b_m(e^{2x},1)} = \frac{\cosh\left(\frac{n}{n+1}x\right)}{\cosh\left(\frac{1}{n+1}x\right)} \times \frac{\cosh\left(\frac{1}{m-1}x\right)}{\cosh\left(\frac{m}{m-1}x\right)}$$

$$= \frac{\cosh((1-\alpha)x)}{\cosh(\alpha x)} \times \frac{\cosh(\beta x)}{\cosh((\beta+1)x)}$$

$$= \frac{\cosh((1-\alpha+\beta)x) + \cosh((1-\alpha-\beta)x)}{2\cosh(\alpha x)\cosh((\beta+1)x)}.$$

We note

$$1 - \alpha + \beta = \frac{nm+1}{(n+1)(m-1)} > 0 \quad \text{and} \quad 1 - \alpha - \beta = \frac{n(m-2)-1}{(n+1)(m-1)} \geq 0$$

thanks to $n \geq 1$ and $m \geq 3$. Since they are majorized by

$$\beta + 1 = \frac{m}{m-1} = \frac{m(n+1)}{(n+1)(m-1)},$$

the ratio $a_n(e^{2x}, 1)/b_m(e^{2x}, 1)$ (whose value at $x = 0$ is 1) is positive definite.

(ii) In this case, we compute

$$\frac{a_n(e^{2x}, 1)}{b_m(e^{2x}, 1)} = \frac{\sinh((1-\alpha)x)}{\cosh(\alpha x)} \times \frac{\cosh(\beta x)}{\sinh((\beta+1)x)}$$

$$= \frac{\sinh((1-\alpha+\beta)x) + \sinh((1-\alpha-\beta)x)}{2\cosh(\alpha x)\sinh((\beta+1)x)}$$

instead. When $m \geq 4$ (i.e., $m \neq 2$), as in (i) both of $1 - \alpha \pm \beta$ are positive and majorized by $\beta + 1$ so that we have the positive definiteness as above. If $m = 2$, then $1 - \alpha - \beta = -(n+1)^{-1} < 0$ so that the above argument does not work. However, since $\beta = 1$ in this case, we have

$$\frac{a_n(e^{2x}, 1)}{b_2(e^{2x}, 1)} = \frac{\sinh((1-\alpha)x)}{\cosh(\alpha x)} \times \frac{\cosh(x)}{\sinh(2x)} = \frac{\sinh((1-\alpha)x)}{2\cosh(\alpha x)\sinh(x)},$$

which is also positive definite thanks to $0 < 1 - \alpha = \frac{n}{n+1} < 1$. Therefore, the ratio $a_n(e^{2x}, 1)/b_m(e^{2x}, 1)$ is always positive definite, and we have

$$\frac{1-\alpha}{\beta+1} = \frac{n}{n+1} \times \frac{m-1}{m}$$

as the value at $x = 0$.   □

The difference version of the Heinz inequality (see [36] and also §6.3, **1**) states

$$|||H^\theta X K^{1-\theta} - H^{1-\theta} X K^\theta|||$$
$$\leq |2\theta - 1| \times |||HX - XK||| \quad \text{(for } \theta \in [0,1]). \tag{8.7}$$

Theorem 8.3, (ii) with $n = m = 2$ means

$$|||H^{\frac{1}{3}} X K^{\frac{2}{3}} - H^{\frac{2}{3}} X K^{\frac{1}{3}}||| \leq \frac{1}{3} |||HX - XK|||,$$

which is exactly (8.7) for the special value $\theta = \frac{1}{3}$. More generally, the theorem (with $m = 2$) states

$$|||\mathbf{A}(2n)||| \leq \frac{n}{2n+1} \times |||HX - XK|||.$$

If one breaks the alternating sum $\mathbf{A}(2n)$ into pieces, then the repeated use of (8.7) (together with the triangle inequality) gives us the constant

$$\sum_{k=1}^{n} \left| 2 \times \frac{k}{2n+1} - 1 \right| = \sum_{k=1}^{n} \frac{2k-1}{2n+1} = \frac{n^2}{2n+1}$$

so that the constant we obtained is far better.

In the next §8.3 we will see that both of

$$n \mapsto |||\mathbf{A}(2n-1)||| \quad \text{and} \quad n \mapsto \frac{2n+1}{2n} \times |||\mathbf{A}(2n)|||$$

are monotone increasing (see Proposition 8.8) so that we have the following uniform lower bounds:

$$|||H^{\frac{1}{2}}XK^{\frac{1}{2}}||| = |||\mathbf{A}(1)||| \leq |||\mathbf{A}(2n-1)|||,$$
$$\frac{3n}{2n+1} \times |||H^{\frac{1}{3}}XK^{\frac{2}{3}} - H^{\frac{2}{3}}XK^{\frac{1}{3}}||| = \frac{2n}{2n+1} \times \frac{3}{2} \times |||\mathbf{A}(2)||| \leq |||\mathbf{A}(2n)|||.$$

On the other hand, from Theorem 8.3 we have the upper bounds

$$|||\mathbf{A}(2n-1)||| \leq |||\mathbf{B}(3)||| = |||HX + XK - H^{\frac{1}{2}}XK^{\frac{1}{2}}|||,$$
$$|||\mathbf{A}(2n)||| \leq |||\mathbf{B}(2)||| = |||HX - XK|||,$$

that can be improved as is seen shortly (the remark below and Theorem 8.5).

Remark 8.4. The bound $|||\mathbf{B}(3)|||$ is comparable with $|||HX + XK|||$:

$$\frac{1}{2}|||HX + XK||| \leq |||HX + XK - H^{\frac{1}{2}}XK^{\frac{1}{2}}||| \leq \frac{3}{2}|||HX + XK|||.$$

The arithmetic-geometric mean inequality (see (1.8)) actually shows the second inequality, and the constant $\frac{3}{2}$ can be removed for the Hilbert-Schmidt norm $|||\cdot||| = ||\cdot||_2$. Indeed, the ratio $(2\cosh(x) - 1)/2\cosh(x)$ is majorized by 1 (see [39, proposition 1.2]). But, since it is not positive definite, the constant $\frac{3}{2}$ (for general unitarily invariant norms) seems optimal. On the other hand, we estimate

$$|||HX + XK||| \leq |||HX + XK - H^{\frac{1}{2}}XK^{\frac{1}{2}}||| + |||H^{\frac{1}{2}}XK^{\frac{1}{2}}|||$$
$$\leq |||HX + XK - H^{\frac{1}{2}}XK^{\frac{1}{2}}||| + \frac{1}{2}|||HX + XK|||.$$

Thus, by subtracting $\frac{1}{2}|||HX + XK|||$ from the both sides, we get the first inequality.

We actually have

Theorem 8.5.

(i) *For each $n = 1, 2, 3, \cdots$ we have*

$$|||\mathbf{A}(2n-1)||| \leq \frac{1}{2}|||HX + XK|||.$$

(ii) *For each $n = 1, 2, 3, \cdots$ we have*

$$|||\mathbf{A}(2n)||| \leq \frac{n}{2n+1} \times |||HX - XK||| \quad \left(\leq \frac{1}{2}|||HX - XK|||\right).$$

Proof. The arithmetic mean $\frac{1}{2}(HX + HK)$ corresponds to $M_2(s, t) = \frac{1}{2}(s+t)$ (see §5.1) and we have

$$\frac{a_{2n-1}(e^{2x}, 1)}{M_2(e^{2x}, 1)} = \frac{\cosh\left(\frac{2n-1}{2n}x\right)}{\cosh(x)\cosh\left(\frac{1}{2n}x\right)}.$$

On the other hand, we compute

$$\frac{a_{2n}(s, t)}{-s+t} = \frac{a_{2n}(s, t)}{(st)^{\frac{1}{2}}\left(\left(\frac{s}{t}\right)^{-\frac{1}{2}} - \left(\frac{s}{t}\right)^{\frac{1}{2}}\right)}$$

$$= \frac{\left(\frac{s}{t}\right)^{-\frac{1}{2} \times \frac{2n}{2n+1}} - \left(\frac{s}{t}\right)^{\frac{1}{2} \times \frac{2n}{2n+1}}}{\left(\left(\frac{s}{t}\right)^{-\frac{1}{2}} - \left(\frac{s}{t}\right)^{\frac{1}{2}}\right)\left(\left(\frac{s}{t}\right)^{-\frac{1}{2} \times \frac{1}{2n+1}} + \left(\frac{s}{t}\right)^{\frac{1}{2} \times \frac{1}{2n+1}}\right)}.$$

Therefore, the corresponding function (i.e., $s = e^{2x}$ and $t = 1$) is

$$\frac{a_{2n}(e^{2x}, 1)}{-e^{2x} + 1} = \frac{\sinh\left(\frac{2n}{2n+1}x\right)}{2\sinh(x)\cosh\left(\frac{1}{2n+1}x\right)}.$$

The two functions are obviously positive definite so that we have the desired inequalities. Notice that the coefficient $\frac{n}{2n+1}$ in (ii) appears as the value of the second function at $x = 0$. □

Let us try to estimate $|||\mathbf{A}(2n-1)|||, |||\mathbf{A}(2n)|||$ (from above and below) in terms of the norms of the "leading terms"

$$H^{\frac{1}{2n}}XK^{\frac{2n-1}{2n}} + H^{\frac{2n-1}{2n}}XK^{\frac{1}{2n}}, \quad H^{\frac{1}{2n+1}}XK^{\frac{2n}{2n+1}} - H^{\frac{2n}{2n+1}}XK^{\frac{1}{2n+1}}.$$

For instance the repeated use of the Heinz inequality (1.3) yields

$$|||\mathbf{A}(5)||| = |||H^{\frac{1}{6}}XK^{\frac{5}{6}} - H^{\frac{2}{6}}XK^{\frac{4}{6}} + H^{\frac{3}{6}}XK^{\frac{3}{6}} - H^{\frac{4}{6}}XK^{\frac{2}{6}} + H^{\frac{5}{6}}XK^{\frac{1}{6}}|||$$

$$\leq |||H^{\frac{1}{6}}XK^{\frac{5}{6}} + H^{\frac{5}{6}}XK^{\frac{1}{6}}||| + |||H^{\frac{2}{6}}XK^{\frac{4}{6}} + H^{\frac{4}{6}}XK^{\frac{2}{6}}||| + |||H^{\frac{3}{6}}XK^{\frac{3}{6}}|||$$

$$\leq \frac{5}{2}|||H^{\frac{1}{6}}XK^{\frac{5}{6}} + H^{\frac{5}{6}}XK^{\frac{1}{6}}|||.$$

Note that this type of reasoning gives us only

$$|||\mathbf{A}(2n-1)||| \leq \frac{2n-1}{2} \times |||H^{\frac{1}{2n}}XK^{\frac{2n-1}{2n}} + H^{\frac{2n-1}{2n}}XK^{\frac{1}{2n}}|||,$$

where the constant $\frac{2n-1}{2}$ blows up. Instead, we actually have

Proposition 8.6.

(i) *We have*

$$\frac{1}{2} \leq \frac{|||\mathbf{A}(2n-1)|||}{|||H^{\frac{1}{2n}}XK^{\frac{2n-1}{2n}} + H^{\frac{2n-1}{2n}}XK^{\frac{1}{2n}}|||} \leq \frac{5}{2}$$

for each n.

(ii) *We have*

$$\frac{n}{2n-1} \leq \frac{|||\mathbf{A}(2n)|||}{|||H^{\frac{1}{2n+1}}XK^{\frac{2n}{2n+1}} - H^{\frac{2n}{2n+1}}XK^{\frac{1}{2n+1}}|||} \leq \frac{3n-2}{2n-1}$$

for each n.

Proof. (i) Note that the sum $\frac{1}{2}\left(H^{\frac{1}{2n}}XK^{\frac{2n-1}{2n}} + H^{\frac{2n-1}{2n}}XK^{\frac{1}{2n}}\right)$ is the Heinz-type mean $A_{\frac{1}{2n}}(H,K)X$ (see (6.1)) and we have

$$A_{\frac{1}{2n}}(s,t) = \frac{1}{2}\left(s^{\frac{1}{2n}}t^{\frac{2n-1}{2n}} + s^{\frac{2n-1}{2n}}t^{\frac{1}{2n}}\right)$$

$$= \frac{(st)^{\frac{1}{2}}}{2} \times \left(\left(\frac{s}{t}\right)^{\frac{1}{2}\times\frac{n-1}{n}} + \left(\frac{s}{t}\right)^{-\frac{1}{2}\times\frac{n-1}{n}}\right).$$

Since

$$\frac{A_{\frac{1}{2n}}(e^{2x},1)}{a_{2n-1}(e^{2x},1)} = \frac{\cosh\left(\frac{1}{2n}x\right)\cosh\left(\frac{n-1}{n}x\right)}{\cosh\left(\frac{2n-1}{2n}x\right)} = \frac{1}{2}\left(1 + \frac{\cosh\left(\frac{2n-3}{2n}x\right)}{\cosh\left(\frac{2n-1}{2n}x\right)}\right)$$

is positive definite, we get the first inequality. To see the second estimate, we need to look at the reciprocal

$$\frac{a_{2n-1}(e^{2x},1)}{A_{\frac{1}{2n}}(e^{2x},1)} = \frac{\cosh\left(\frac{2n-1}{2n}x\right)}{\cosh\left(\frac{1}{2n}x\right)\cosh\left(\frac{n-1}{n}x\right)}$$

and Lemma 8.1 says the desired inequality.

(ii) To see the first inequality, we have to look at the ratio

$$\frac{\cosh\left(\frac{1}{2n+1}x\right)\sinh\left(\frac{2n-1}{2n+1}x\right)}{\sinh\left(\frac{2n}{2n+1}x\right)} = \frac{1}{2}\left(1 + \frac{\sinh\left(\frac{2n-2}{2n+1}x\right)}{\sinh\left(\frac{2n}{2n+1}x\right)}\right).$$

This is positive definite and the value at $x = 0$ is

$$\frac{1}{2}\left(1+\frac{2n-2}{2n}\right)=\frac{2n-1}{2n}.$$

The reciprocal

$$\frac{\sinh\left(\frac{2n}{2n+1}x\right)}{\cosh\left(\frac{1}{2n+1}x\right)\sinh\left(\frac{2n-1}{2n+1}x\right)}$$

is equal to

$$\frac{\sinh\left(\frac{1}{2n+1}x\right)\cosh\left(\frac{2n-1}{2n+1}x\right)+\cosh\left(\frac{1}{2n+1}x\right)\sinh\left(\frac{2n-1}{2n+1}x\right)}{\cosh\left(\frac{1}{2n+1}x\right)\sinh\left(\frac{2n-1}{2n+1}x\right)}$$

$$=2+\frac{\sinh\left(\frac{1}{2n+1}x\right)\cosh\left(\frac{2n-1}{2n+1}x\right)-\cosh\left(\frac{1}{2n+1}x\right)\sinh\left(\frac{2n-1}{2n+1}x\right)}{\cosh\left(\frac{1}{2n+1}x\right)\sinh\left(\frac{2n-1}{2n+1}x\right)}$$

$$=2-\frac{\sinh\left(\frac{2n-2}{2n+1}x\right)}{\cosh\left(\frac{1}{2n+1}x\right)\sinh\left(\frac{2n-1}{2n+1}x\right)}.$$

Note that the subtracted ratio in the last expression is positive definite with the value $\frac{2n-2}{2n-1}$ at $x=0$. Thus, the whole function can be expressed as the Fourier transform of a signed measure with total variation at most

$$2+\frac{2n-2}{2n-1}=\frac{2(3n-2)}{2n-1},$$

showing the second inequality.   □

We next try to obtain uniform (upper and lower) bounds for $|||\mathbf{B}(m)|||$'s, and begin with the case $m=3,5,7,\cdots$ (odd). At first we note

$$\frac{M_2(s,t)}{b_m(s,t)}=\frac{\left(\frac{s}{t}\right)^{\frac{1}{2}}+\left(\frac{s}{t}\right)^{-\frac{1}{2}}}{2}\times\frac{\left(\frac{s}{t}\right)^{\frac{1}{2}\times\frac{1}{m-1}}+\left(\frac{s}{t}\right)^{-\frac{1}{2}\times\frac{1}{m-1}}}{\left(\frac{s}{t}\right)^{\frac{1}{2}\times\frac{m}{m-1}}+\left(\frac{s}{t}\right)^{-\frac{1}{2}\times\frac{m}{m-1}}}.$$

Since

$$\frac{M_2(e^{2x},1)}{b_m(e^{2x},1)}=\frac{\cosh(x)\cosh\left(\frac{1}{m-1}x\right)}{\cosh\left(\frac{m}{m-1}x\right)}=\frac{1}{2}\left(1+\frac{\cosh\left(\frac{m-2}{m-1}x\right)}{\cosh\left(\frac{m}{m-1}x\right)}\right)$$

is positive definite, we conclude

$$\frac{1}{2}|||HX+XK|||=|||M_2(H,K)X|||\le|||\mathbf{B}(m)|||\quad(m=3,5,7,\cdots).$$

For an upper bound we obviously have

$$|||\mathbf{B}(m)||| \le |||HX + XK||| + |||\mathbf{A}(m-2)||| \le \frac{3}{2}|||HX + XK|||$$

thanks to (8.1) and Theorem 8.5, (i). (Note that a slightly different estimate based on Proposition 8.6, (i) is also possible.) We however point out that a multiple of the norm $||| \int_0^1 H^x XK^{1-x} dx|||$ of the logarithmic mean cannot majorize $|||\mathbf{B}(m)|||$. Indeed, the leading term of $b_m(s,1)$ is s while we have

$$M_1(s,1) = \int_0^1 s^x dx = \frac{s-1}{\log s}.$$

We next consider the case $m = 2, 4, 6, \cdots$ (even). Theorem 8.5, (ii) and (8.1) give rise to an upper bound for $|||\mathbf{B}(m)|||$ as follows:

$$|||\mathbf{B}(m)||| \le |||HX - XK||| + |||\mathbf{A}(m-2)|||$$
$$\le |||HX - XK||| + \frac{m-2}{m-1} \times \frac{1}{2} \times |||HX - XK|||$$
$$= \left(1 + \frac{m-2}{2(m-1)}\right) \times |||HX - XK|||$$

for $m = 2, 4, 6, \cdots$. To get a lower bound for $|||\mathbf{B}(m)|||$ (with m even), as in the proof of Theorem 8.5 we compute

$$\frac{-e^{2x} + 1}{b_m(e^{2x}, 1)} = \frac{\cosh\left(\frac{1}{m-1}x\right)}{\sinh\left(\frac{m}{m-1}x\right)} \times \sinh(x)$$

$$= \frac{\sinh\left(\left(1 + \frac{1}{m-1}\right)x\right) + \sinh\left(\left(1 - \frac{1}{m-1}\right)x\right)}{2\sinh\left(\frac{m}{m-1}x\right)}$$

$$= \frac{1}{2}\left(1 + \frac{\sinh\left(\frac{m-2}{m-1}x\right)}{\sinh\left(\frac{m}{m-1}x\right)}\right).$$

This function is positive definite (thanks to $\frac{m-2}{m-1} < \frac{m}{m-1}$) with the value $\frac{m-1}{m}$ at $x = 0$ so that we conclude

$$\frac{1}{2}|||HX - XK||| \le \frac{m-1}{m} \times |||\mathbf{B}(m)|||.$$

Summing up the discussions so far we have shown

Theorem 8.7.

(i) *For $m = 3, 5, 7, \cdots$ we have*

$$\frac{1}{2} \le \frac{|||\mathbf{B}(m)|||}{|||HX + XK|||} \le \frac{3}{2}.$$

(ii) *For $m = 2, 4, 6, \cdots$ we have*

$$\left(\frac{1}{2} \le\right) \frac{m}{2(m-1)} \le \frac{|||\mathbf{B}(m)|||}{|||HX - XK|||} \le 1 + \frac{m-2}{2(m-1)} \left(\le \frac{3}{2}\right).$$

8.3 Monotonicity of norms

Monotonicity for $|||\mathbf{A}(n)|||$ and $|||\mathbf{B}(m)|||$ (either odd or even) is studied in this section. We begin with the former (which is quite straight-forward).

Proposition 8.8.

(i)   *The norm $|||\mathbf{A}(2n-1)|||$ is monotone increasing in n ($n = 1, 2, 3, \cdots$).*

(ii) *The quantity $\dfrac{2n+1}{2n} \times |||\mathbf{A}(2n)|||$ is also monotone increasing in n ($n = 1, 2, 3, \cdots$).*

Proof. For $n' \geq n$ (odd) we have

$$\frac{a_n(e^{2x}, 1)}{a_{n'}(e^{2x}, 1)} = \frac{\cosh\left(\frac{n}{n+1}x\right)}{\cosh\left(\frac{1}{n+1}x\right)} \times \frac{\cosh\left(\frac{1}{n'+1}x\right)}{\cosh\left(\frac{n'}{n'+1}x\right)}.$$

Because of $\frac{n}{n+1} \leq 1, \frac{1}{n'+1} \leq \frac{1}{n+1}$ and $\frac{n}{n+1} \leq \frac{n'}{n'+1}$ both functions are positive definite and the values at $x = 0$ are 1. On the other hand, for $n' \geq n$ (even) we have

$$\frac{a_n(e^{2x}, 1)}{a_{n'}(e^{2x}, 1)} = \frac{\sinh\left(\frac{n}{n+1}x\right)}{\cosh\left(\frac{1}{n+1}x\right)} \times \frac{\cosh\left(\frac{1}{n'+1}x\right)}{\sinh\left(\frac{n'}{n'+1}x\right)}.$$

Because of $\frac{1}{n'+1} \leq \frac{1}{n+1}$ and $\frac{n}{n+1} \leq \frac{n'}{n'+1}$ this function is positive definite and the value at $x = 0$ is $\frac{n'+1}{n'} \times \frac{n}{n+1}$. $\square$

The case $|||\mathbf{B}(m)|||$ is more involved, and monotone decreasingness is obtained only in a weak sense (except for the Hilbert-Schmidt norm $\|\cdot\|_2$).

Theorem 8.9.

(i)   *We have*

$$|||\mathbf{B}(2m+3) - \mathbf{B}(2m+1)||| \leq |||\mathbf{B}(2m+1)|||$$

for $m = 1, 2, 3, \cdots$, and in particular

$$|||\mathbf{B}(2m+3)||| \leq 2|||\mathbf{B}(2m+1)|||.$$

(ii) *We have*

$$|||\mathbf{B}(2m+2) - \mathbf{B}(2m)||| \leq \frac{2m-1}{2m} \times |||\mathbf{B}(2m)|||$$

for $m = 1, 2, 3, \cdots$, and in particular

$$|||\mathbf{B}(2m+2)||| \leq \left(1 + \frac{2m-1}{2m}\right) \times |||\mathbf{B}(2m)||| \left(\leq 2 \times |||\mathbf{B}(2m)|||\right).$$

(iii) *For the Hilbert-Schmidt norm* $\|\cdot\|_2$ *we have the monotone decreasingness*

$$\|\mathbf{B}(2m+3)\|_2 \leq \|\mathbf{B}(2m+1)\|_2 \quad and \quad \|\mathbf{B}(2m+2)\|_2 \leq \|\mathbf{B}(2m)\|_2.$$

(iv) *The monotone decreasingness*

$$|||\mathbf{B}(m')||| \leq |||\mathbf{B}(m)||| \quad (for\ m' > m\ odd)$$

fails to hold for general unitarily invariant norms.

Proof. (i) For $m' \geq m \geq 3$ (odd) we have

$$\frac{b_{m'}(e^{2x},1)}{b_m(e^{2x},1)} = \frac{\cosh\left(\frac{1}{m-1}x\right)}{\cosh\left(\frac{m}{m-1}x\right)} \times \frac{\cosh\left(\frac{m'}{m'-1}x\right)}{\cosh\left(\frac{1}{m'-1}x\right)}. \tag{8.8}$$

For convenience we set

$$\alpha = \frac{1}{m'-1} \quad and \quad \beta = \frac{1}{m-1}.$$

We compute

$$\frac{b_{m'}(e^{2x},1)}{b_m(e^{2x},1)} - 1 = \frac{\cosh((\alpha+1)x)\cosh(\beta x)}{\cosh((\beta+1)x)\cosh(\alpha x)} - 1$$

$$= \frac{\cosh((\alpha+1)x)\cosh(\beta x) - \cosh((\beta+1)x)\cosh(\alpha x)}{\cosh((\beta+1)x)\cosh(\alpha x)}$$

$$= \frac{1}{\cosh((\beta+1)x)\cosh(\alpha x)} \times$$

$$\left\{ \Big(\cosh(\alpha x)\cosh(x) + \sinh(\alpha x)\sinh(x)\Big)\cosh(\beta x) \right.$$

$$\left. - \Big(\cosh(\beta x)\cosh(x) + \sinh(\beta x)\sinh(x)\Big)\cosh(\alpha x)\right\}$$

$$= \frac{\sinh(x)}{\cosh((\beta+1)x)\cosh(\alpha x)} \times \Big\{\sinh(\alpha x)\cosh(\beta x) - \cosh(\alpha x)\sinh(\beta x)\Big\}$$

$$= \frac{\sinh(x)\sinh((\alpha-\beta)x)}{\cosh((\beta+1)x)\cosh(\alpha x)}.$$

Since $m' \geq m$, i.e., $0 < \alpha \leq \beta$, the above last quantity is negative so that we have the (point-wise) monotone decreasingness

$$(0 \leq)\ b_{m'}(s,t) \leq b_m(s,t),$$

showing (iii) in the odd case (see [39, Proposition 1.2]). Notice

$$\frac{b_{m'}(e^{2x},1)}{b_m(e^{2x},1)} = 1 + \frac{\sinh(x)\sinh((\alpha-\beta)x)}{\cosh((\beta+1)x)\cosh(\alpha x)}$$

$$= 1 + \frac{\cosh((\alpha-\beta+1)x) - \cosh((\alpha-\beta-1)x)}{2\cosh((\beta+1)x)\cosh(\alpha x)}. \tag{8.9}$$

We now assume $m' = m + 2$ (i.e., $\alpha = \frac{1}{m+1}$ and $\beta = \frac{1}{m-1}$) so that

$$\begin{cases} \alpha - \beta + 1 = \frac{m^2-3}{(m+1)(m-1)} > 0, \\[2mm] \alpha - \beta - 1 = -\frac{m^2+1}{(m+1)(m-1)} < 0. \end{cases} \tag{8.10}$$

Notice that the hyperbolic cosine function is even and

$$\frac{m^2-3}{(m+1)(m-1)} \leq \frac{m}{m-1} \quad \text{and} \quad \frac{m^2+1}{(m+1)(m-1)} \leq \frac{m}{m-1} \tag{8.11}$$

with $\dfrac{m}{m-1} = \beta + 1$. Consequently, the second term in the far right side of (8.9) is a difference of two positive definite functions (with the value $\frac{1}{2}$ at $x = 0$), showing (i).

(ii) For $m' \geq m \geq 2$ (even) we have

$$\frac{b_{m'}(e^{2x}, 1)}{b_m(e^{2x}, 1)} = \frac{\cosh\left(\frac{1}{m-1}x\right)}{\sinh\left(\frac{m}{m-1}x\right)} \times \frac{\sinh\left(\frac{m'}{m'-1}x\right)}{\cosh\left(\frac{1}{m'-1}x\right)} = \frac{\sinh((\alpha+1)x)\cosh(\beta x)}{\sinh((\beta+1)x)\cosh(\alpha x)}$$

instead with α and β appearing above. Hence, the computations in (i) are changed as follows:

$$\frac{b_{m'}(e^{2x}, 1)}{b_m(e^{2x}, 1)} - 1 = \frac{\sinh((\alpha+1)x)\cosh(\beta x) - \sinh((\beta+1)x)\cosh(\alpha x)}{\sinh((\beta+1)x)\cosh(\alpha x)}$$

$$= \frac{1}{\sinh((\beta+1)x)\cosh(\alpha x)} \times$$

$$\left\{ \Big(\sinh(\alpha x)\cosh(x) + \cosh(\alpha x)\sinh(x)\Big)\cosh(\beta x) \right.$$

$$\left. - \Big(\sinh(\beta x)\cosh(x) + \cosh(\beta x)\sinh(x)\Big)\cosh(\alpha x)\right\}$$

$$= \frac{\cosh(x)}{\sinh((\beta+1)x)\cosh(\alpha x)} \times \Big\{\sinh(\alpha x)\cosh(\beta x) - \cosh(\alpha x)\sinh(\beta x)\Big\}$$

$$= \frac{\cosh(x)\sinh((\alpha-\beta)x)}{\sinh((\beta+1)x)\cosh(\alpha x)}.$$

Since $0 < \alpha \leq \beta$, the above last quantity is negative so that once again we have the point-wise monotone decreasingness

$$|b_{m'}(s, t)| \leq |b_m(s, t)|,$$

showing (iii) in the even case. We have

$$\frac{b_{m'}(e^{2x}, 1)}{b_m(e^{2x}, 1)} = 1 + \frac{\cosh(x)\sinh((\alpha-\beta)x)}{\sinh((\beta+1)x)\cosh(\alpha x)}$$

$$= 1 + \frac{\sinh((\alpha-\beta+1)x) + \sinh((\alpha-\beta-1)x)}{2\sinh((\beta+1)x)\cosh(\alpha x)}.$$

We now assume $m' = m + 2$ as before. Since $\alpha - \beta - 1$ is negative (see (8.10)) and the hyperbolic sine function is odd, we have

$$\frac{b_{m+2}(e^{2x}, 1)}{b_m(e^{2x}, 1)} = 1 + \frac{\sinh((\alpha - \beta + 1)x) - \sinh((-\alpha + \beta + 1)x)}{2\sinh((\beta + 1)x)\cosh(\alpha x)}.$$

Therefore, (8.11) once again yields that the above ratio is a difference of positive definite functions. Note that their values at $x = 0$ are

$$\frac{\alpha - \beta + 1}{\beta + 1} = \frac{m^2 - 3}{(m+1)(m-1)} \times \frac{1}{2} \times \frac{m-1}{m} = \frac{m^2 - 3}{2m(m+1)},$$

$$\frac{-\alpha + \beta + 1}{\beta + 1} = \frac{m^2 + 1}{(m+1)(m-1)} \times \frac{1}{2} \times \frac{m-1}{m} = \frac{m^2 + 1}{2m(m+1)}$$

respectively. They sum up to $\frac{m-1}{m}$ so that (by changing (even) m to $2m$) we get the inequality in (ii).

(iv)   When m is odd, it is obvious that the function $b_m(s, t)$ in $s, t > 0$ is a symmetric homogeneous function such that $b_m(s, s) = s$ for all $s > 0$. Although $b_m(s, 1)$ is not non-decreasing in s, the proof of (ii) $\Rightarrow$ (iv) in Theorem 3.7 (i.e., (ii) $\Rightarrow$ (v) in [39, Theorem 1.1]) works well (see also the proof of Theorem A.3 in §A.1). Thus, if $|||\mathbf{B}(m')||| \leq |||\mathbf{B}(m)|||$ (for odd $m' > m \geq 3$) were valid for all unitarily invariant norms, then

$$f(x) = \frac{b_{m'}(e^{2x}, 1)}{b_m(e^{2x}, 1)}$$

would be a positive definite function, i.e., $f(x) = \hat{\nu}(x)$ for some probability measure ν (because of $f(0) = 1$). However, by Proposition A.7 and (8.8) we would have

$$\nu(\{0\}) = \lim_{x \to \pm\infty} f(x) = 1,$$

meaning $f(x) = 1$, a contradiction. $\square$

In the part (iv) of the theorem, the monotone decreasingness $|||B(m')||| \leq |||B(m)|||$ (for $m' > m$ odd) actually fails to hold for the operator norm $||| \cdot ||| = \| \cdot \|$ and for the trace norm $||| \cdot ||| = \| \cdot \|_1$. Indeed, the proof of [39, Theorem 1.1] says that if the decreasingness (in case of matrices) were valid for one of these norms then we would have the positive definiteness of the above function $f(x)$.

We are unable to determine what happens in the even case.

8.4 Notes and references

Trivial modification of the argument for the proof of the first inequality in Remark 8.4 enables us to obtain

$$\frac{2+x}{2}|||HX+XK||| \le |||HX+XK+xH^{1/2}XK^{1/2}|||$$

for $x \in (-2, 0]$. This fact and the ordinary Heinz inequality (1.3) imply that the inequality (i) in §3.7, **2** holds true for each $\theta \in [0,1]$ as long as $x \in (-2,0]$. Inequalities involving the norm of an operator of the form

$$H^{\theta}XK^{1-\theta} + H^{1-\theta}XK^{\theta} + xH^{\frac{1}{2}}XK^{\frac{1}{2}}$$

have been studied by many authors (see [13, 78, 83] for instance). Note that the cases $\theta = \frac{3}{4}, 1$ (and $x = -1$) correspond to $\mathbf{A}(3), \mathbf{B}(3)$ respectively, and quite thorough investigation on inequalities involving these quantities will be carried out in the forthcoming article [55].

Note that the logarithmic-geometric mean inequality (see (1.8)) says

$$|||\mathbf{A}(1)||| = |||H^{\frac{1}{2}}XK^{\frac{1}{2}}||| \le |||\int_0^1 H^x XK^{1-x}dx|||,$$

which should be compared with Theorem 8.5, (i). The estimate of this form is no longer valid for $|||\mathbf{A}(3)|||$, but it is possible to estimate (more generally) $|||\mathbf{A}(2n-1)|||$ by a constant multiple of $|||\int_0^1 H^x XK^{1-x}dx|||$ ([55]).

A

Appendices

We collect six appendices here. In §A.1 we will deal with certain non-symmetric means (by weakening the axioms stated in Definition 3.1), and we will see that all the results in §3.2 remain valid for such means (sometimes with obvious modification). In §A.2–A.6 some technical results used in the main body of the monograph are clarified.

A.1 Non-symmetric means

We can deal with a wider class of (not necessarily symmetric) homogeneous means for positive scalars. We denote by $\widetilde{\mathfrak{M}}$ the set of all continuous positive real functions $M(s,t)$ for $s,t > 0$ satisfying

$$\begin{cases} \text{the properties (b), (c) in Definition 3.1,} \\ \text{and } M(s,s) = s \text{ for } s > 0 \text{ in place of (d) there.} \end{cases}$$

For $M, N \in \widetilde{\mathfrak{M}}$ the order $M \preceq N$ is introduced in the same way as in Definition 3.2, that is, $M \preceq N$ if and only if there exists a symmetric measure ν on $\mathbf{R}$ such that $M(e^x, 1) = \hat{\nu}(x) N(e^x, 1)$ $(x \in \mathbf{R})$.

Remark A.1. Here are some remarks on the above measure ν.

(i) The measure ν in Definition 3.2 was automatically symmetric (since so are $M(s,t)$ and $N(s,t)$) while it is now a part of the requirement.

(ii) When $M \preceq N$, we have

$$M(s,t)/N(s,t) = M(t,s)/N(t,s)$$

(although $M(s,t)$ and $N(s,t)$ might be asymmetric). In fact, since ν is symmetric, we compute

$$M(s,t)/N(s,t) = M(s/t,1)/N(s/t,1) = \hat{\nu}(\log s - \log t)$$
$$= \hat{\nu}(\log t - \log s) = M(t/s,1)/N(t/s,1) = M(t,s)/N(t,s).$$

(iii) The measure ν is a probability measure because of

$$M(e^0, 1) = N(e^0, 1) = 1.$$

As in the case of a mean in $\mathfrak{M}$, the domain of $M \in \widetilde{\mathfrak{M}}$ extends to $[0, \infty) \times [0, \infty)$ as follows:

$$M(s, 0) = \lim_{t \searrow 0} M(s, t) = sM(1, 0) \qquad (s > 0),$$

$$M(0, t) = \lim_{s \searrow 0} M(s, t) = tM(0, 1) \qquad (t > 0),$$

and $M(0, 0) = 0$ while $M(1, 0) \neq M(0, 1)$ in general. So, for positive operators $H, K \in B(\mathcal{H})$ one can define the double integral transformation $M(H, K)X$ first for $X \in \mathcal{C}_2(\mathcal{H})$ and then for all $X \in B(\mathcal{H})$ whenever M is a Schur multiplier relative to (H, K).

We will show that the main results in §3.2 remain valid also for means in $\widetilde{\mathfrak{M}}$, and we begin with generalizations of Theorem 3.4 and Corollary 3.5 (see also Proposition 8.2).

Theorem A.2. *Assume that means M, N in $\widetilde{\mathfrak{M}}$ satisfy $M \preceq N$.*

(i) *The integral expressions (i.e., (3.8) and (3.9)) in Theorem 3.4 remain valid with the modification of (3.8) by*

$$M(H, K)X = \int_{-\infty}^{\infty} (Hs_H)^{ix}(N(H, K)X)(Ks_K)^{-ix} d\nu(x)$$
$$+ M(1, 0)HX(1 - s_K) + M(0, 1)(1 - s_H)XK.$$

(ii) *The norm inequality in Corollary 3.5 also holds true.*

Proof. The proof of Proposition 8.2 works here thanks to (8.3) and Remark A.1, (ii). Note that the estimate (8.6) there is not necessary since we have the stronger estimate $M(1, 0) \leq N(1, 0)$ (due to Remark A.1, (iii)) as in the proof of Theorem 3.4. Of course (ii) follows from Theorem A.5 as usual.   □

We are now ready to prove a generalization of Theorem 3.7.

Theorem A.3. *The conditions (i)–(iv) in Theorem 3.7 are all equivalent for means M, N in $\widetilde{\mathfrak{M}}$.*

Proof. Theorem A.2, (i) and (ii) guarantee (iv) $\Rightarrow$ (i) and (iv) $\Rightarrow$ (ii) respectively. The proof of (iv) $\Rightarrow$ (iii) is the same as in the proof of Theorem 3.7 while (i) $\Rightarrow$ (iv) is trivial as in the proof of Theorem 3.7.

It remains show (ii) $\Rightarrow$ (iv) and (iii) $\Rightarrow$ (iv). To this end, it suffices to prove $M \preceq N$ under the assumption that (iii) holds for all matrices $H \geq 0$ and X of any size. Now, for any $s_1, \ldots, s_n > 0$ put $H = \mathrm{diag}(s_1, \ldots, s_n)$. Since (iii) means

$$\|[M(s_i, s_j)] \circ X\| \le \|[N(s_i, s_j)] \circ X\|$$

for all $n \times n$ matrices X, one gets $\|T \circ X\| \le \|X\|$ with $T = \left[\frac{M(s_i,s_j)}{N(s_i,s_j)}\right]_{i,j=1,\cdots,n}$.
Since $\mathrm{Tr}((T \circ X)Y) = \mathrm{Tr}(X(T^t \circ Y))$ for all $n \times n$ matrices X, Y, one has
$\|T^t \circ Y\|_1 \le \|Y\|_1$ so that

$$\|T \circ Y\|_1 = \|(T \circ Y)^t\|_1 = \|T^t \circ Y^t\|_1 \le \|Y^t\|_1 = \|Y\|_1.$$

Choose the matrix of all entries 1 for Y; then the above estimate gives

$$\|T\|_1 = \|T \circ Y\| \le \|Y\|_1 = n.$$

On the other hand, since $M(s,s) = N(s,s) = s$, the diagonals of T are all 1
and consequently

$$\|T\|_1 \ge \mathrm{Tr}\, T = n.$$

Hence we have seen $\|T\|_1 = \mathrm{Tr}\, T$. Let $T = V|T|$ with a unitary matrix V, and
assume that $|T|$ is diagonalized with a unitary matrix U as follows:

$$|T| = U\mathrm{diag}(\lambda_1, \ldots, \lambda_n)U^*.$$

Then, we observe

$$\sum_{i=1}^{n} \lambda_i = \|T\|_1 = \mathrm{Tr}\, T = \mathrm{Tr}\big(U^*VU\mathrm{diag}(\lambda_1, \ldots, \lambda_n)\big) = \sum_{i=1}^{n} \lambda_i u_{ii}$$

with the unitary matrix $U^*VU = [u_{ij}]$. Note $u_{ii} = 1$ as long as $\lambda_i > 0$ (thanks
to the obvious facts $|u_{ii}| \le 1$ and $\lambda_i \ge 0$). Hence, by assuming say

$$\lambda_1, \ldots, \lambda_k > 0 = \lambda_{k+1} = \lambda_{k+2} = \cdots = \lambda_n,$$

we can write

$$U^*VU = I_k \oplus W_{n-k},$$

and consequently

$$T = U(I_k \oplus W_{n-k})U^*U\mathrm{diag}(\lambda_1, \ldots, \lambda_k, 0, \ldots, 0)U^*$$
$$= U\mathrm{diag}(\lambda_1, \ldots, \lambda_k, 0, \ldots, 0)U^*.$$

This means $T = |T| \ge 0$, and $M \preceq N$ is shown. $\square$

Let us present two simple examples for which Theorem A.2 is useful. Firstly
let us assume

$$0 < \alpha \le \beta < 1, \quad 0 \le \delta \le \min\{\alpha, 1 - \beta\},$$

and we set

$$M(s,t) = s^\alpha t^{1-\alpha} + s^\beta t^{1-\beta},$$
$$N(s,t) = s^{\alpha-\delta} t^{1-\alpha+\delta} + s^{\beta+\delta} t^{1-\beta-\delta}.$$

Although M, N fail to be symmetric, $\frac{1}{2}M, \frac{1}{2}N$ fall into $\widetilde{\mathfrak{M}}$ and they satisfy

$$\frac{M(e^x,1)}{N(e^x,1)} = \frac{e^{\alpha x} + e^{\beta x}}{e^{(\alpha-\delta)x} + e^{(\beta+\delta)x}}$$

$$= \frac{e^{-\frac{\beta-\alpha}{2}x} + e^{\frac{\beta-\alpha}{2}x}}{e^{-(\frac{\beta-\alpha}{2}+\delta)x} + e^{(\frac{\beta-\alpha}{2}+\delta)x}} = \frac{\cosh\left(\frac{\beta-\alpha}{2}x\right)}{\cosh\left((\frac{\beta-\alpha}{2}+\delta)x\right)},$$

which is positive definite (see [39, (1.5)] for example). Therefore, we have $M \preceq N$, and Theorem A.2, (ii) implies

$$|||H^\alpha X K^{1-\alpha} + H^\beta X K^{1-\beta}||| \le |||H^{\alpha-\delta} X B^{1-\alpha+\delta} + A^{\beta+\delta} X K^{1-\beta-\delta}|||$$

for all unitarily invariant norms and all operators $H, K \ge 0$ and X. It is also possible to derive this inequality from Heinz-type inequalities (see Chapter 6), and details are left to the reader.

Secondly we assume

$$0 < \alpha_1, \ldots, \alpha_k < 1 \quad \text{and} \quad 0 < \beta < \min\{\alpha_1, \ldots, \alpha_k, 1-\alpha_1, \ldots, 1-\alpha_k\}.$$

For $\lambda_1, \ldots, \lambda_k \ge 0$ with $\sum_{i=1}^k \lambda_i = 1$ we consider $M, N \in \widetilde{\mathfrak{M}}$ defined by

$$M(s,t) = \sum_{i=1}^k \lambda_i s^{\alpha_i} t^{1-\alpha_i},$$

$$N(s,t) = \frac{1}{2}\sum_{i=1}^k \lambda_i\left(s^{\alpha_i+\beta}t^{1-\alpha_i-\beta} + s^{\alpha_i-\beta}t^{1-\alpha_i+\beta}\right).$$

Note $N(s,t) = \frac{s^\beta t^{-\beta} + s^{-\beta} t^\beta}{2} \times M(s,t)$ and

$$\frac{M(e^x,1)}{N(e^x,1)} = \frac{1}{\cosh(\beta x)}$$

is positive definite (see Example 3.6, (a)). Thus, once again Theorem A.2, (ii) implies

$$|||\sum_{i=1}^k \lambda_i H^{\alpha_i} X K^{1-\alpha_i}|||$$

$$\le \frac{1}{2}|||\sum_{i=1}^k \lambda_i\left(H^{\alpha_i+\beta} X K^{1-\alpha_i-\beta} + H^{\alpha_i-\beta} X K^{1-\alpha_i+\beta}\right)|||.$$

In particular,

$$|||\lambda H^\alpha X K^{1-\alpha} + (1-\lambda)H^{1-\alpha} X K^\alpha|||$$

$$\le \frac{1}{2}|||\lambda H^{2\alpha-\frac{1}{2}} X K^{\frac{3}{2}-2\alpha} + H^{\frac{1}{2}} X K^{\frac{1}{2}} + (1-\lambda)H^{\frac{3}{2}-2\alpha} X K^{2\alpha-\frac{1}{2}}|||$$

for every $\frac{1}{4} \leq \alpha \leq \frac{3}{4}$ and $0 \leq \lambda \leq 1$.

The equivalent conditions for $M, N \in \widetilde{\mathfrak{M}}$ obtained in Theorem A.3 are somewhat too restrictive, and it is also interesting to characterize the situation where $|||M(H, K)X||| \leq C|||N(H, K)X|||$ holds with some universal constant C (for all $H, K \geq 0$ and X). A sufficient condition is that $M(e^x, 1)/N(e^x, 1) = \hat{\mu}(x)$ $(x \in \mathbf{R})$ for some signed measure μ on $\mathbf{R}$. This condition implies the above inequality with $C = \|\mu\|$ (the total variation of μ). A typical application of this reasoning is the weak Young inequality (6.4) whose full details were worked out in [54]. Note that this method was employed in Chapter 8 (although $a_n(s, t), b_m(s, t)$ there need not fall into $\widetilde{\mathfrak{M}}$).

A.2 Norm inequality for operator integrals

We assume that $F : \Omega \to B(\mathcal{H})$ is a weakly measurable operator-valued function on a measure space (Ω, μ) in the sense that the function $x \in \Omega \mapsto (F(x)\xi, \eta)$ is measurable for each vectors $\xi, \eta \in \mathcal{H}$. In this section the operator integral

$$\int_\Omega F(x) \, d\mu(x)$$

is considered, and its (unitarily invariant) norm estimate will be studied.

The proof of the next lemma is based on the separability assumption on the ambient Hilbert space $\mathcal{H}$.

Lemma A.4. *For each unitarily invariant norm $||| \cdot |||$, the function*

$$x \in \Omega \mapsto |||F(x)||| \in [0, \infty]$$

is measurable.

Proof. At first we claim that $x \mapsto \mu_n(F(x))$ is measurable for each $n = 0, 1, \ldots$, where $\mu_n(\cdot)$ denotes the n-th singular number. When $n = 0$, we note $\mu_0(F(x)) = \|F(x)\|$, i.e., the operator norm, and by choosing a dense sequence $\{\xi_i\}_{i=1,2,\ldots}$ in the unit ball of $\mathcal{H}$ we have

$$\|F(x)\| = \sup_{i,j} |(F(x)\xi_i, \xi_j)|.$$

Therefore, the weak measurability guarantees the measurability of $x \mapsto \mu_0(F(x))$. To deal with general n's, we recall the famous trick appearing for example in the proof of the Weyl inequality (see [77, §1, (v)] for details) based on anti-symmetric tensors. The main ingredient of the trick is the fact that the n-fold anti-symmetric tensor product $\wedge^n(F(x)) \in B(\wedge^n \mathcal{H})$ satisfies

$$\| \wedge^n (F(x))\| = \prod_{k=0}^{n-1} \mu_k(F(x)).$$

Hence, the preceding argument (using a dense sequence) applied for $\wedge^n(F(x))$ guarantees the measurability of $x \mapsto \prod_{k=0}^{n-1} \mu_k(F(x))$ (for each n) and we are done.

Let Φ be the symmetric norm for (finite) sequences corresponding to $|||\cdot|||$. Since

$$|||F(x)||| = \lim_{n \to \infty} \Phi(\mu_0(F(x)), \mu_1(F(x)), \dots, \mu_n(F(x)), 0, 0, \dots),$$

to prove the lemma it suffices to check the measurability of

$$x \in \Omega \mapsto \Phi(\mu_0(F(x)), \mu_1(F(x)), \dots, \mu_n(F(x)), 0, 0, \dots) \in [0, \infty)$$

for each fixed n. Note that this map is the composition of the measurable map $x \mapsto (\mu_0(F(x)), \mu_1(F(x)), \dots, \mu_n(F(x)))$ (thanks to the first half of the proof) followed by

$$(a_0, a_1, \dots, a_n) \in \mathbf{R}^{n+1} \mapsto \Phi(a_0, a_1, \dots, a_n, 0, 0, \dots) \in [0, \infty).$$

However, the latter is a norm and hence continuous so that the composition is clearly measurable. □

Next, we further require that a weakly measurable operator-valued function $F : \Omega \to B(\mathcal{H})$ satisfies the $\|\cdot\|$-integrability

$$\int_\Omega \|F(x)\| \, d\mu(x) < \infty.$$

Then, the operator integral $Z = \int_\Omega F(x) d\mu(x) \in B(\mathcal{H})$ can be defined in the weak sense, i.e.,

$$(Z\xi, \eta) = \int_\Omega (F(x)\xi, \eta) \, d\mu(x) \quad (\text{for } \xi, \eta \in \mathcal{H}),$$

and the following estimate is straight-forward:

$$\|Z\| \le \int_\Omega \|F(x)\| \, d\mu(x).$$

The next theorem asserts that a similar norm estimate remains valid for every unitarily invariant norm.

Theorem A.5. *Let $|||\cdot|||$ be a unitarily invariant norm, and we assume that a weakly measurable operator-valued function $F : \Omega \to B(\mathcal{H})$ on a measure space (Ω, μ) satisfies the $\|\cdot\|$-integrability*

$$\int_\Omega \|F(x)\| \, d\mu(x) < \infty.$$

Then, the norm of the operator $Z = \int_\Omega F(x) d\mu(x) \in B(\mathcal{H})$ (defined in the weak sense as above) admits the following estimate:

$$|||Z||| \le \int_\Omega |||F(x)||| \, d\mu(x) \ (\le \infty).$$

Proof. We at first point out that one can reduce the proof to the case where (Ω, μ) is a finite measure and $|||F(\cdot)|||$ is bounded.

(i) We can assume $\mu(\Omega) < \infty$. Indeed, (Ω, μ) can be assumed to be σ-finite because F is supported on a σ-finite measurable set. So let $\{\Omega_i\}_{i=1,2,\cdots}$ be an increasing sequence of measurable subsets with $\mu(\Omega_i) < \infty$ (for each i) exhausting the whole space Ω. We set

$$Z_i = \int_{\Omega_i} F(x)\, d\mu(x) \quad \text{(in the weak sense)}.$$

Then, the $\|\cdot\|$-integrability of $F(\cdot)$ implies

$$|\,((Z - Z_i)\xi, \eta)\,| \leq \|\xi\| \times \|\eta\| \times \int_{\Omega \setminus \Omega_i} \|F(x)\|\, d\mu(x) \quad (\xi, \eta \in \mathcal{H}),$$

which tends to 0 as $i \to \infty$ due to the Lebesgue dominated convergence theorem, i.e., $\{Z_i\}_{i=1,2,\cdots}$ tends to Z in the weak operator topology. Therefore, if the result is known for Ω_i's (of finite measure), then by the lower semi-continuity of $|||\cdot|||$ in the weak operator topology (see [37, Proposition 2.11]) we get

$$|||Z||| \leq \liminf_{i \to \infty} |||Z_i||| \leq \liminf_{i \to \infty} \int_{\Omega_i} |||F(x)|||\, d\mu(x) = \int_{\Omega} |||F(x)|||\, d\mu(x).$$

Here, the last equality follows from the monotone convergence theorem.

(ii) We can assume the $|||\cdot|||$-boundedness of F. Indeed, if $\int_{\Omega} |||F(x)|||d\mu(x) = \infty$, we have nothing to prove. Hence, we may and do assume the integrability of $|||F(\cdot)|||$. In particular, we have $|||F(x)||| < \infty$ for μ-a.e. x. We set

$$\tilde{\Omega}_n = \{x \in \Omega : |||F(x)||| \leq n\} \text{ and } \tilde{Z}_n = \int_{\tilde{\Omega}_n} F(x)\, d\mu(x) \text{ (in the weak sense)}.$$

Then $\{\tilde{\Omega}_n\}_{n=1,2,\cdots}$ is increasing with $\bigcup_n \tilde{\Omega}_n = \Omega$ (up to a null set). The same arguments as in (i) show that $\{\tilde{Z}_n\}_{n=1,2,\cdots}$ tends to Z in the weak operator topology, and we have $|||Z||| \leq \int_{\Omega} |||F(x)|||d\mu(x)$ (if the result is known for $\tilde{\Omega}_n$'s).

Thanks to (i) and (ii), we can assume $\mu(\Omega) < \infty$ and the $|||\cdot|||$-boundedness of F in the rest of the proof. We choose and fix $\varepsilon > 0$ and α satisfying $\alpha < |||Z|||$. ($|||Z|||$ could be ∞ a priori, in which case α can be anything. However, our arguments in what follows will rule out the possibility of $|||Z||| = \infty$.)

The set $\{X \in B(\mathcal{H}) : |||X||| > \alpha\}$ is an open neighborhood of Z relative to the weak topology from the lower semi-continuity of $|||\cdot|||$. Hence, vectors $\xi_1, \xi_2, \ldots, \xi_N \in \mathcal{H}$ and $\delta > 0$ can be chosen in such a way that

$$|\,((X - Z)\xi_s, \xi_t)\,| \leq \delta \quad (s, t = 1, \ldots, N) \quad \Longrightarrow \quad |||X||| > \alpha. \tag{A.1}$$

Choose and fix a pair $(s, t) \in \{1, 2, \ldots, N\}^2$ for a moment. Since $\|\cdot\|$ is majorized by $|||\cdot|||$, $(F(\cdot)\xi_s, \xi_t)$ is a bounded measurable function. By dividing the range of the function into small pieces and considering the corresponding preimages, one can choose a finite measurable partition $\{S_1, S_2, \ldots, S_\ell\}$ of Ω such that $|(F(x)\xi_s, \xi_t) - (F(x')\xi_s, \xi_t)| \leq \frac{\delta}{\mu(\Omega)}$ if x, x' belong to the same S_i. Note that we have finitely many (s,t)'s and do the same for each of (s,t)'s. By considering the common refinement of all the partitions obtained in this procedure (the refinement is denoted by $\{S_1, S_2, \ldots, S_\ell\}$ again), we conclude

$$|(F(x)\xi_s, \xi_t) - (F(x')\xi_s, \xi_t)| \leq \frac{\delta}{\mu(\Omega)} \quad \text{(for all } s,t) \tag{A.2}$$

as long as x, x' sit in the same S_i $(i = 1, 2, \ldots, \ell)$.

On the other hand, since $\||F(\cdot)|\|$ is bounded, we can also take a finite measurable partition $\{T_1, T_2, \ldots, T_m\}$ of Ω such that

$$\sum_{j=1}^{m} M_j \mu(T_j) \leq \int_\Omega \||F(x)|\| \, d\mu(x) + \varepsilon \tag{A.3}$$

with

$$M_j = \sup\{\||F(x)|\| : x \in T_j\} \quad (j = 1, 2, \ldots, m).$$

Let $\{Q_k\}_{k=1,2,\cdots,n}$ be a renumbering of $\{S_i \cap T_j\}_{i=1,2,\cdots,\ell; j=1,2,\cdots,m}$, and we choose x_k from each Q_k $(k = 1, 2, \ldots, n)$. Being a refinement of $\{S_i\}_{i=1,2,\cdots,\ell}$, the property (A.2) remains valid for the Q_k's. Firstly, for each s, t we estimate

$$\left| \left(\left(\sum_{k=1}^{n} F(x_k)\mu(Q_k) - Z \right) \xi_s, \xi_t \right) \right| = \left| \sum_{k=1}^{n} \int_{Q_k} ((F(x_k) - F(x))\xi_s, \xi_t) \, d\mu(x) \right|$$

$$\leq \sum_{k=1}^{n} \int_{Q_k} |((F(x_k) - F(x))\xi_s, \xi_t)| \, d\mu(x)$$

$$\leq \frac{\delta}{\mu(\Omega)} \sum_{k=1}^{n} \mu(Q_k) = \delta.$$

This means that

$$X = \sum_{k=1}^{n} F(x_k)\mu(Q_k) \in B(\mathcal{H})$$

satisfies the assumption of (A.1), and consequently we get $\||X|\| > \alpha$. Secondly, from the above definition of M_j we observe

$$\||X|\| = \left\| \left| \sum_{k=1}^{n} F(x_k)\mu(Q_k) \right| \right\| \leq \sum_{k=1}^{n} \||F(x_k)|\| \mu(Q_k) \leq \sum_{j=1}^{m} M_j \mu(T_j)$$

since $\{Q_k\}_{k=1,2,\cdots,n}$ is a refinement of $\{T_j\}_{j=1,2,\cdots,m}$. This estimate and (A.3) imply

$$\||X|\| \leq \int_\Omega \||F(x)|\| \, d\mu(x) + \varepsilon.$$

Therefore, we conclude

$$\alpha < |||X||| \leq \int_\Omega |||F(x)||| \, d\mu(x) + \varepsilon.$$

Since $\alpha \ (< |||Z|||)$ and $\varepsilon \ (> 0)$ were arbitrary, we are done. $\square$

The following proof based on the duality $\mathcal{I}_{|||\cdot|||} = \left(\mathcal{I}^{(0)}_{|||\cdot|||'}\right)^*$ (see Remark 4.2, (4) and the first part of the proof below) is also worth pointing out:

Alternative proof of Theorem A.5. Let $||| \cdot |||'$ be the conjugate norm of $||| \cdot |||$, and recall that the duality $\mathcal{I}_{|||\cdot|||} = \left(\mathcal{I}^{(0)}_{|||\cdot|||'}\right)^*$ is given by the bilinear form $(X, Y) \in \mathcal{I}_{|||\cdot|||} \times \mathcal{I}^{(0)}_{|||\cdot|||'} \mapsto \mathrm{Tr}(XY) \in \mathbf{C}$. On the other hand, from the definition of the separable ideal $\mathcal{I}^{(0)}_{|||\cdot|||'}$, each $Y \in \mathcal{I}^{(0)}_{|||\cdot|||'}$ can be approximated by finite-rank operators with norm at most $|||Y|||'$. Therefore, we have

$$|||X||| = \sup\{|\mathrm{Tr}(XY)| : Y \text{ is of finite-rank and } |||Y|||' \leq 1\}$$

(see the proof of [37, Proposition 2.11]). For each $Y = \sum_{i=1}^n \xi_i \otimes \eta_i^c$ with $|||Y|||' \leq 1$ we estimate

$$|\mathrm{Tr}(ZY)| = \left| \sum_{i=1}^n (Z\xi_i, \eta_i) \right| = \left| \int_\Omega \sum_{i=1}^n (F(x)\xi_i, \eta_i) \, d\mu(x) \right|$$

$$\leq \int_\Omega \left| \sum_{i=1}^n (F(x)\xi_i, \eta_i) \right| d\mu(x) = \int_\Omega \left| \mathrm{Tr}(F(x)Y)) \right| d\mu(x)$$

$$\leq \int_\Omega |||F(x)||| \, d\mu(x).$$

Thus, by taking the supremum over Y's, we get the conclusion. $\square$

A.3 Decomposition of max$\{s,t\}$

We assume that the integral operator T acting on $L^2([a,b])$ with a kernel $k(s,t)$ $(\in L^2([a,b] \times [a,b]))$ is positive (i.e., $k(s,t)$ is a positive definite in the sense of §3.4), and let $\{\lambda_n\}_{n=1,2,\cdots}$ be the (strictly) positive eigenvalues $\lambda_1 \geq \lambda_2 \geq \lambda_3 \geq \cdots > 0$ (with multiplicities counted). The spectral decomposition theorem states $T = \sum_n \lambda_n \phi_n \otimes \phi_n^c$ for an orthonormal system $\{\phi_n(t)\}_{n=1,2,\cdots}$ $(\subseteq L^2([a,b]))$ of corresponding eigenvectors. The following result (that is a consequence of Dini's theorem) is known as Mercer's theorem (see [79, Theorem 7.7.2], [81, p. 125] or [82, Chapter 3 §2 32]): If a positive definite kernel $k(s,t)$ is a continuous function on $[a,b] \times [a,b]$, then so are eigenfunctions $\phi_n(t)$ and moreover we have

$$k(s,t) = \sum_n \lambda_n \phi_n(s)\overline{\phi_n(t)},$$

the series being uniformly and absolutely convergent on $[a,b] \times [a,b]$. Based on this theorem one can prove the absolute convergence

$$\min\{s,t\} = 2\sum_{n=1}^{\infty} \left(\frac{2}{(2n-1)\pi}\right)^2 \sin\left(\frac{(2n-1)\pi s}{2}\right) \sin\left(\frac{(2n-1)\pi t}{2}\right) \quad (\text{A.4})$$

for $(s,t) \in [0,1] \times [0,1]$ (see the end of the section), which plays an important role in analysis of the Brownian process. With slightly more involved arguments the next decomposition can be also obtained.

Theorem A.6. *The function* $\max\{s,t\}$ *on* $[0,1] \times [0,1]$ *admits the absolutely convergent decomposition*

$$\max\{s,t\} = 2\Big(\frac{\alpha^2 - 1}{\alpha^4} \times \cosh(\alpha s)\cosh(\alpha t)$$

$$-\sum_{n=1}^{\infty} \frac{1 + \alpha_n^2}{\alpha_n^4} \times \cos(\alpha_n s)\cos(\alpha_n t)\Big).$$

Here, α (> 1) *is a unique positive real satisfying* $\tanh(\alpha) - \frac{1}{\alpha} = 0$ *while* $\alpha_1 < \alpha_2 < \cdots$ *are the positive roots for the equation* $\tan(x) + \frac{1}{x} = 0$.

Proof. We consider the integral operator with the kernel $\max\{s,t\}$ acting on the Hilbert space $L^2([0,1]; dt)$, which is a self-adjoint operator sitting in $\mathcal{C}_2(L^2([0,1]))$. Let $x(t)$ be an eigenvector with an eigenvalue $\lambda \in \mathbf{R}$:

$$\lambda x(t) = \int_0^1 \max\{t,s\}x(s)\,ds = t\int_0^t x(s)\,ds + \int_t^1 sx(s)\,ds. \quad (\text{A.5})$$

When $\lambda = 0$, the differentiation of the right-hand side gives us

$$(0 =) \int_0^t x(s)\,ds + tx(t) - tx(t) = \int_0^t x(s)\,ds.$$

Hence, we must have $x(s) = 0$, that is, the operator is non-singular. In the rest let us assume $\lambda \neq 0$. Because of

$$\int_0^t x(s)\,ds = \lambda x'(t)$$

we observe $x'(0) = 0$ and $x(t) = \lambda x''(t)$.

We begin with the case $\lambda > 0$. The general solution for the differential equation $x'' - \lambda^{-1}x = 0$ is

$$x(t) = A\exp\left(\lambda^{-\frac{1}{2}}t\right) + B\exp\left(-\lambda^{-\frac{1}{2}}t\right).$$

However, the boundary condition $x'(0) = 0$ forces $A = B$ so that an eigenvector must be a constant multiple of

$$x(t) = \cosh\left(\lambda^{-\frac{1}{2}}t\right).$$

The direct computation of the right side of (A.5) with this function yields

$$t \int_0^t \cosh\left(\lambda^{-\frac{1}{2}}s\right) ds + \int_t^1 s \cosh\left(\lambda^{-\frac{1}{2}}s\right) ds$$
$$= \lambda^{\frac{1}{2}} \sinh\left(\lambda^{-\frac{1}{2}}\right) - \lambda \cosh\left(\lambda^{-\frac{1}{2}}\right) + \lambda \cosh\left(\lambda^{-\frac{1}{2}}t\right). \qquad (A.6)$$

Therefore, $x(t)$ is an eigenvector if and only if

$$\sinh\left(\lambda^{-\frac{1}{2}}\right) - \lambda^{\frac{1}{2}} \cosh\left(\lambda^{-\frac{1}{2}}\right) = 0, \text{ i.e., } \lambda^{-\frac{1}{2}} = \alpha,$$

showing that $\lambda = 1/\alpha^2$ is the only positive eigenvalue. The square of the L^2-norm of the eigenvector $x(t) = \cosh(\alpha t)$ is

$$\int_0^1 \cosh^2(\alpha s) \, ds = \frac{1}{2} \int_0^1 (1 + \cosh(2\alpha s)) \, ds$$
$$= \frac{1}{2}\left(1 + \frac{\sinh(2\alpha)}{2\alpha}\right) = \frac{1}{2}\left(1 + \frac{\sinh(\alpha)\cosh(\alpha)}{\alpha}\right).$$

We note

$$\sinh(\alpha)\cosh(\alpha) = \tanh(\alpha)\cosh^2(\alpha) = \frac{\tanh(\alpha)}{1 - \tanh^2(\alpha)}$$

so that the above quantity is equal to

$$\frac{1}{2}\left(1 + \frac{1}{\alpha} \times \frac{\frac{1}{\alpha}}{1 - \left(\frac{1}{\alpha}\right)^2}\right) = \frac{\alpha^2}{2(\alpha^2 - 1)}.$$

Therefore, a unit eigenvector (for the eigenvalue $\lambda = 1/\alpha^2$) is given by

$$x_0(t) = \frac{\sqrt{2}\sqrt{\alpha^2 - 1}}{\alpha} \times \cosh(\alpha t).$$

We next move to the case $\lambda < 0$. By setting $\tilde{\lambda} = -\lambda > 0$, we consider the differential equation $x'' + \tilde{\lambda}^{-1}x = 0$ with the general solution

$$A \sin\left(\tilde{\lambda}^{-\frac{1}{2}}t\right) + B \cos\left(\tilde{\lambda}^{-\frac{1}{2}}t\right).$$

As before the boundary condition $x'(0) = 0$ forces $A = 0$ and we set

$$x(t) = \cos\left(\tilde{\lambda}^{-\frac{1}{2}}t\right).$$

Note that the computation (A.6) is replaced by

$$t \int_0^t \cos\left(\tilde{\lambda}^{-\frac{1}{2}} s\right) ds + \int_t^1 s \cos\left(\tilde{\lambda}^{-\frac{1}{2}} s\right) ds$$
$$= \tilde{\lambda}^{\frac{1}{2}} \sin\left(\tilde{\lambda}^{-\frac{1}{2}}\right) + \tilde{\lambda} \cos\left(\tilde{\lambda}^{-\frac{1}{2}}\right) - \tilde{\lambda} \cos\left(\tilde{\lambda}^{-\frac{1}{2}} t\right).$$

Therefore, $x(t)$ is an eigenvector if and only if

$$\sin\left(\tilde{\lambda}^{-\frac{1}{2}}\right) + \tilde{\lambda}^{\frac{1}{2}} \cos\left(\tilde{\lambda}^{-\frac{1}{2}}\right) = 0,$$

that is, $\tilde{\lambda}^{-\frac{1}{2}}$ must be a (positive) solution for $\tan(x) + \frac{1}{x} = 0$. We assume $\tilde{\lambda}^{-\frac{1}{2}} = \alpha_n$ $(n = 1, 2, \cdots)$. This means that $\lambda = -\tilde{\lambda} = -1/\alpha_n^2$ is a negative eigenvalue with an eigenvector $x(t) = \cos(\alpha_n t)$. The preceding computations for normalization should be modified in the following way:

$$\int_0^1 \cos^2(\alpha_n s)\, ds = \frac{1}{2} \int_0^1 (1 + \cos(2\alpha_n s))\, ds$$
$$= \frac{1}{2}\left(1 + \frac{\sin(2\alpha_n)}{2\alpha_n}\right) = \frac{1}{2}\left(1 + \frac{\sin(\alpha_n)\cos(\alpha_n)}{\alpha_n}\right)$$
$$= \frac{1}{2}\left(1 + \frac{1}{\alpha_n} \times \frac{\tan(\alpha_n)}{1 + \tan^2(\alpha_n)}\right)$$
$$= \frac{1}{2}\left(1 + \frac{1}{\alpha_n} \times \frac{-\frac{1}{\alpha_n}}{1 + \left(-\frac{1}{\alpha_n}\right)^2}\right) = \frac{\alpha_n^2}{2(1 + \alpha_n^2)}.$$

Thus, we conclude that

$$x_n(t) = \frac{\sqrt{2}\sqrt{1 + \alpha_n^2}}{\alpha_n} \times \cos(\alpha_n t) \quad (n = 1, 2, \cdots)$$

is a normalized eigenvector for the negative eigenvalue $\lambda = -1/\alpha_n^2$.

The arguments so far show that the integral operator T with the kernel $\max\{s, t\}$ admits the spectral decomposition

$$T = \frac{1}{\alpha^2} x_0 \otimes x_0^c - \sum_{n=1}^{\infty} \frac{1}{\alpha_n^2} x_n \otimes x_n^c.$$

Since the difference $\frac{1}{\alpha^2} x_0 \otimes x_0^c - T$ is a positive integral operator with the continuous kernel

$$\frac{1}{\alpha^2} x_0(s)\overline{x_0(t)} - \max\{s, t\}$$
$$= \frac{1}{\alpha^2} \times \frac{2(\alpha^2 - 1)}{\alpha^2} \times \cosh(\alpha s)\cosh(\alpha t) - \max\{s, t\},$$

the desired convergence follows from Mercer's theorem. $\quad\square$

Assume $0 \le H, K \le 1$ for instance. Then, the above theorem permits the following alternative definition:

$$M_\infty(H, K)X = 2\Big(\frac{\alpha^2 - 1}{\alpha^4} \times \cosh(\alpha H)X \cosh(\alpha K)$$
$$- \sum_{n=1}^{\infty} \frac{1 + \alpha_n^2}{\alpha_n^4} \times \cos(\alpha_n H)X \cos(\alpha_n K)\Big), \quad (A.7)$$

which coincides with the one considered in previous chapters (see Remark 2.5, (ii)). Substitutions $s = t = 0$ and $s = t = 1$ to the series in the theorem give rise to

$$\frac{\alpha^2 - 1}{\alpha^4} = \sum_{n=1}^{\infty} \frac{1 + \alpha_n^2}{\alpha_n^4}, \quad (A.8)$$

$$\frac{\alpha^2 - 1}{\alpha^4} \times \cosh^2 \alpha = \frac{1}{2} + \sum_{n=1}^{\infty} \frac{1 + \alpha_n^2}{\alpha_n^4} \times \cos^2 \alpha_n. \quad (A.9)$$

The expression (A.7) clearly shows $|||M_\infty(H, K)H||| \le \kappa |||X|||$ with

$$\kappa = 2\left(\frac{\alpha^2 - 1}{\alpha^4} \times \cosh^2 \alpha + \sum_{n=1}^{\infty} \frac{1 + \alpha_n^2}{\alpha_n^4}\right).$$

Note that (A.8) and $\tanh \alpha = 1/\alpha$ yield

$$\kappa = 2 \times \frac{\alpha^2 - 1}{\alpha^4} \times (\cosh^2 \alpha + 1) = 2 \times \frac{2\alpha^2 - 1}{\alpha^4}$$

while (A.9) and $\tan \alpha_n = -1/\alpha_n$ show

$$\kappa = 2\left(\frac{1}{2} + \sum_{n=1}^{\infty} \frac{1 + \alpha_n^2}{\alpha_n^4} \times (\cos^2 \alpha_n + 1)\right) = 1 + 2\sum_{n=1}^{\infty} \frac{2\alpha_n^2 + 1}{\alpha_n^4}.$$

From the first expression for κ and $\alpha > 1$ we observe $\kappa < 2$. On the other hand, the second and the obvious fact $\alpha_n < n\pi$ (for $n = 1, 2, \cdots$) imply

$$\kappa > 1 + 2\sum_{n=1}^{\infty} \frac{2\pi^2 n^2 + 1}{\pi^4 n^4} = 1 + \frac{4}{\pi^2}\sum_{n=1}^{\infty} \frac{1}{n^2} + \frac{2}{\pi^4}\sum_{n=1}^{\infty} \frac{1}{n^4} = 1 + \frac{31}{45}$$

(thanks to $\sum_{n=1}^{\infty} n^{-2} = \pi^2/6$ and $\sum_{n=1}^{\infty} n^{-4} = \pi^4/90$). Hence, (although the expression (A.7) makes it trivial that max$\{s, t\}$ is a Schur multiplier) it seems impossible to get the optimal constant $\frac{2}{\sqrt{3}}$ obtained in Theorem 3.12.

Both of positive and negative eigenvalues appeared in the proof of Theorem A.6. This phenomenon corresponds to the fact that M_∞ is not majorized (in the sense of Definition 3.2) by the geometric mean $G = M_{1/2}$ (see Proposition 3.10). The proof for (A.4) is easier since all the eigenvalues (which are actually $(2/(2n - 1)\pi)^2$ with $n = 1, 2, \cdots$) are positive due to $M_{-\infty} \preceq G$. Details are left to the reader as an easy exercise.

A.4 Cesàro limit of the Fourier transform

In this section the formula (i.e., Proposition A.9) that have appeared before Theorem 3.4 and some related results are explained.

Proposition A.7. *For every complex measure μ on $\mathbf{R}$, we have*

$$\mu(\{0\}) = \lim_{T \to \infty} \frac{1}{2T} \int_{-T}^{T} \hat{\mu}(t) \, dt \, .$$

Proof. To prove the proposition, it suffices to show

$$\mu(\{0\}) = 0 \implies \lim_{T \to \infty} \frac{1}{2T} \int_{-T}^{T} \hat{\mu}(t) \, dt = 0$$

(by considering $\mu - \mu(\{0\})\delta_0$), and hence let us assume $\mu(\{0\}) = 0$. The Fubini theorem shows

$$\frac{1}{2T} \int_{-T}^{T} \hat{\mu}(t) \, dt = \frac{1}{2T} \int_{-T}^{T} \left(\int_{-\infty}^{\infty} e^{ist} \, d\mu(s) \right) dt$$

$$= \int_{-\infty}^{\infty} \left(\frac{1}{2T} \int_{-T}^{T} e^{ist} \, dt \right) d\mu(s)$$

$$= \int_{-\infty}^{\infty} \frac{1}{2T} \times \frac{e^{isT} - e^{-isT}}{is} \, d\mu(s)$$

$$= \int_{-\infty}^{\infty} \frac{\sin(sT)}{sT} \, d\mu(s).$$

Therefore, for each (small) $\delta > 0$ we estimate

$$\left| \frac{1}{2T} \int_{-T}^{T} \hat{\mu}(t) \, dt \right| \leq \int_{|s| < \delta} \left| \frac{\sin(sT)}{sT} \right| d|\mu|(s) + \int_{|s| \geq \delta} \left| \frac{\sin(sT)}{sT} \right| d|\mu|(s)$$

$$\leq |\mu|((-\delta, \delta)) + \frac{1}{\delta T} |\mu|(\mathbf{R}) \, ,$$

and hence

$$\limsup_{T \to \infty} \left| \frac{1}{2T} \int_{-T}^{T} \hat{\mu}(t) \, dt \right| \leq |\mu|((-\delta, \delta)) \, .$$

Note that the assumption $\mu(\{0\}) = 0$ implies $|\mu|(\{0\}) = 0$. Since $|\mu|(\mathbf{R}) < \infty$, we have $|\mu|((-\delta, \delta)) \to 0$ as $\delta \to +0$ and consequently

$$\limsup_{T \to \infty} \left| \frac{1}{2T} \int_{-T}^{T} \hat{\mu}(t) \, dt \right| = 0$$

as desired.   □

If $\hat{\mu}(t) \to \alpha$ as $t \to \pm\infty$, then it is plain to observe

$$\lim_{T \to \infty} \frac{1}{2T} \int_{-T}^{T} \hat{\mu}(t)\, dt = \alpha,$$

and hence we have

Corollary A.8. *If* $\lim\limits_{|t| \to \infty} \hat{\mu}(t)$ *exists, then it is equal to* $\mu(\{0\})$.

Here is the formula mentioned before Theorem 3.4.

Proposition A.9. *For a complex measure* μ *on* $\mathbf{R}$ *we have*

$$\sum_{t \in \mathbf{R}} |\mu(\{t\})|^2 = \lim_{T \to \infty} \frac{1}{2T} \int_{-T}^{T} |\hat{\mu}(t)|^2\, dt\,.$$

Proof. Let us set $\tilde{\mu}(S) = \overline{\mu(-S)}$ for $S \subset \mathbf{R}$. Then we easily observe $\hat{\tilde{\mu}}(t) = \overline{\hat{\mu}(t)}$ and hence

$$\widehat{\mu * \tilde{\mu}}(t) = \hat{\mu}(t)\hat{\tilde{\mu}}(t) = |\hat{\mu}(t)|^2.$$

On the other hand, we note

$$\mu * \tilde{\mu}(\{0\}) = \int_{-\infty}^{\infty} \tilde{\mu}(\{-t\})\, d\mu(t) = \int_{-\infty}^{\infty} \overline{\mu(\{t\})}\, d\mu(t) = \sum_{t \in \mathbf{R}} |\mu(\{t\})|^2\,.$$

Hence, Proposition A.7 applied to $\mu * \tilde{\mu}$ gives the result. □

A.5 Reflexivity and separability of operator ideals

Here the reflexivity and separability of symmetrically normed ideals are discussed, and we need the following general facts on Banach spaces:

(i)   The uniform convexity implies the reflexivity ([74, Chapter V, Problem 15]).
(ii) A Banach space X is reflexive if and only if so is the dual X^* ([25, Corollary II.3.24]).
(iii) If X^* is separable, then so is X ([74, Theorem III.7]).

Proposition A.10. *Let* $|||\cdot|||$ *be a unitarily invariant norm. If either* $\mathcal{I}_{|||\cdot|||}$ *or* $\mathcal{I}_{|||\cdot|||}^{(0)}$ *is reflexive, then* $\mathcal{I}_{|||\cdot|||}$ *is separable, i.e.,* $\mathcal{I}_{|||\cdot|||} = \mathcal{I}_{|||\cdot|||}^{(0)}$ *(see* [29, §III.6]).

Proof. We assume that $\mathcal{I}_{|||\cdot|||}$ is reflexive, and let $||| \cdot |||'$ be the conjugate norm of $||| \cdot |||$. Then, the general duality $\mathcal{I}_{|||\cdot|||} = \left(\mathcal{I}_{|||\cdot|||'}^{(0)} \right)^*$ and (ii) yield the reflexivity of $\mathcal{I}_{|||\cdot|||'}^{(0)}$, so that we have

$$\left(\mathcal{I}_{|||\cdot|||} \right)^* = \left(\mathcal{I}_{|||\cdot|||'}^{(0)} \right)^{**} = \mathcal{I}_{|||\cdot|||'}^{(0)}.$$

Hence $\left(\mathcal{I}_{|||\cdot|||}\right)^{*}$ is separable and so is $\mathcal{I}_{|||\cdot|||}$ by (iii).

Next, we assume that $\mathcal{I}^{(0)}_{|||\cdot|||}$ is reflexive. Since the dual space $\mathcal{I}_{|||\cdot|||'} = \left(\mathcal{I}^{(0)}_{|||\cdot|||}\right)^{*}$ is reflexive by (ii), the first half of the proof (applied to $||| \cdot |||'$) guarantees that $\mathcal{I}_{|||\cdot|||'} = \mathcal{I}^{(0)}_{|||\cdot|||'}$, so that we observe

$$\mathcal{I}_{|||\cdot|||} = \left(\mathcal{I}^{(0)}_{|||\cdot|||'}\right)^{*} = \left(\mathcal{I}_{|||\cdot|||'}\right)^{*} = \left(\mathcal{I}^{(0)}_{|||\cdot|||}\right)^{**} = \mathcal{I}^{(0)}_{|||\cdot|||},$$

showing the separability of $\mathcal{I}_{|||\cdot|||}$. $\square$

Corollary A.11. *If one of $\mathcal{I}_{|||\cdot|||}$, $\mathcal{I}^{(0)}_{|||\cdot|||}$, $\mathcal{I}_{|||\cdot|||'}$ and $\mathcal{I}^{(0)}_{|||\cdot|||'}$ is reflexive, then all of them are reflexive and we have the separability $\mathcal{I}_{|||\cdot|||} = \mathcal{I}^{(0)}_{|||\cdot|||}$, $\mathcal{I}_{|||\cdot|||'} = \mathcal{I}^{(0)}_{|||\cdot|||'}$. We also get the same conclusion when one of $\mathcal{I}_{|||\cdot|||}$, $\mathcal{I}^{(0)}_{|||\cdot|||}$, $\mathcal{I}_{|||\cdot|||'}$ and $\mathcal{I}^{(0)}_{|||\cdot|||'}$ is uniformly convex.*

Proof. The proof of Proposition A.10 actually shows

$$\begin{cases} \text{the reflexivity of } \mathcal{I}_{|||\cdot|||} \Rightarrow \text{the reflexivity of } \mathcal{I}^{(0)}_{|||\cdot|||'}, \\ \qquad\qquad\qquad\qquad\qquad\quad \text{and the separability of } \mathcal{I}_{|||\cdot|||}, \\ \text{the reflexivity of } \mathcal{I}^{(0)}_{|||\cdot|||} \Rightarrow \text{the reflexivity of } \mathcal{I}_{|||\cdot|||'} \\ \qquad\qquad\qquad\qquad\qquad\quad \text{and the separability of } \mathcal{I}_{|||\cdot|||}. \end{cases}$$

Application of these to $||| \cdot |||$ and $||| \cdot |||'$ easily shows the first statement while the second statement follows from the first and (i). $\square$

A.6 Fourier transform of $1/\cosh^{\alpha}(t)$

The Fourier transform of $1/\cosh^{\alpha}(t)$ for $\alpha > 0$ can be found in standard tables of Fourier transforms (see [65, p. 33] for instance). However, the authors are unable to find details in the literature so that computations are given here.

Since the function in question is even, we note

$$I = \int_{-\infty}^{\infty} \frac{1}{\cosh^{\alpha}(t)} e^{ist} dt = \int_{-\infty}^{\infty} \frac{\cos(st)}{\cosh^{\alpha}(t)} dt = 2^{\alpha} \int_{-\infty}^{\infty} \frac{\cos(st)}{(e^{t} + e^{-t})^{\alpha}} dt.$$

The change of variables $t = \frac{1}{2} \log\left(\frac{x}{1-x}\right)$ (hence $\frac{x}{1-x} = e^{2t}$ and $dt = \frac{dx}{2x(1-x)}$) gives us

$$I = 2^{\alpha-1} \int_{0}^{1} \frac{\cos\left(\frac{s}{2} \log\left(\frac{x}{1-x}\right)\right)}{\left(\sqrt{\frac{x}{1-x}} + \sqrt{\frac{1-x}{x}}\right)^{\alpha}} \times \frac{dx}{x(1-x)}$$

$$= 2^{\alpha-1} \int_{0}^{1} (x(1-x))^{\frac{\alpha}{2}} \times \cos\left(\frac{s}{2} \log\left(\frac{x}{1-x}\right)\right) \times \frac{dx}{x(1-x)}$$

$$= 2^{\alpha-1} \int_{0}^{1} x^{\frac{\alpha}{2}-1} (1-x)^{\frac{\alpha}{2}-1} \cos\left(\frac{s}{2} \log\left(\frac{x}{1-x}\right)\right) dx.$$

Notice

$$
\begin{cases}
\cos\!\left(\tfrac{s}{2}\log\!\left(\tfrac{x}{1-x}\right)\right) = \cos\!\left(\tfrac{s}{2}\log x\right)\cos\!\left(\tfrac{s}{2}\log(1-x)\right) \\
\qquad\qquad\qquad + \sin\!\left(\tfrac{s}{2}\log x\right)\sin\!\left(\tfrac{s}{2}\log(1-x)\right), \\[2mm]
x^z = x^{\operatorname{Re} z}\cos(\operatorname{Im} z\,\log x) + ix^{\operatorname{Re} z}\sin(\operatorname{Im} z\,\log x) \quad \text{(for } x > 0).
\end{cases}
$$

Based on these we easily observe

$$
x^{\frac{\alpha}{2}-1}(1-x)^{\frac{\alpha}{2}-1}\cos\!\left(\tfrac{s}{2}\log\!\left(\tfrac{x}{1-x}\right)\right) = \operatorname{Re}\!\left(x^{\frac{\alpha}{2}-1+\frac{is}{2}}(1-x)^{\frac{\alpha}{2}-1-\frac{is}{2}}\right),
$$

and consequently

$$
I = 2^{\alpha-1}\operatorname{Re}\left(\int_0^1 x^{\frac{\alpha}{2}-1+\frac{is}{2}}(1-x)^{\frac{\alpha}{2}-1-\frac{is}{2}}\,dx\right).
$$

The integral here is

$$
B\!\left(\frac{\alpha}{2}+\frac{is}{2},\ \frac{\alpha}{2}-\frac{is}{2}\right) = \frac{\Gamma\!\left(\frac{\alpha}{2}+\frac{is}{2}\right)\Gamma\!\left(\frac{\alpha}{2}-\frac{is}{2}\right)}{\Gamma(\alpha)}
$$

in terms of the B-function (and the Γ-function), showing

$$
I = \frac{2^{\alpha-1}}{\Gamma(\alpha)}\times\operatorname{Re}\!\left(\Gamma\!\left(\frac{\alpha}{2}+\frac{is}{2}\right)\Gamma\!\left(\frac{\alpha}{2}-\frac{is}{2}\right)\right).
$$

Note $\Gamma(\bar z) = \overline{\Gamma(z)}$ by the Schwarz reflection principle so that the above formula actually means

$$
\int_{-\infty}^{\infty}\frac{1}{\cosh^\alpha(t)}\,e^{ist}\,dt = \frac{2^{\alpha-1}}{\Gamma(\alpha)}\times\left|\Gamma\!\left(\frac{\alpha}{2}+\frac{is}{2}\right)\right|^2,
$$

or equivalently,

$$
\int_{-\infty}^{\infty}\frac{1}{\cosh^{1/\alpha}(\alpha t)}\,e^{ist}\,dt = \frac{2^{\frac{1}{\alpha}-1}}{\alpha\Gamma\!\left(\frac{1}{\alpha}\right)}\times\left|\Gamma\!\left(\frac{1+is}{2\alpha}\right)\right|^2. \tag{A.10}
$$

References

1. T. Ando, *Majorizations, doubly stochastic matrices, and comparison of eigenvalues*, Linear Algebra Appl., **118** (1989), 163–248.
2. T. Ando, *Majorizations and inequalities in matrix theory*, Linear Algebra Appl., **199** (1994), 17–67.
3. T. Ando, *Matrix Young inequalities*, Oper. Theory Adv. Appl., **75** (1995), 33–38.
4. T. Ando, R. A. Horn and C. R. Johnson, *The singular values of a Hadamard product: a basic inequality*, Linear and Multilinear Algebra, **21** (1987), 345–365.
5. T. Ando and K. Okubo, *Induced norms of the Schur multiplier operator*, Linear Algebra Appl., **147** (1991), 181–199.
6. J. Bergh and J. Löfström, *Interpolation Spaces, An Introduction*, Springer-Verlag, Berlin-New York, 1976.
7. R. Bhatia, *A simple proof of an operator inequality of Jocić and Kittaneh*, J. Operator Theory, **31** (1994), 21–22.
8. R. Bhatia, *Matrix Analysis*, Springer-Verlag, New York, 1996.
9. R. Bhatia, *On the exponential metric increasing property*, preprint (2002).
10. R. Bhatia and C. Davis, *More matrix forms of the arithmetic geometric mean inequality*, SIAM J. Matrix Anal. Appl., **14** (1993), 132–136.
11. R. Bhatia and C. Davis, *A Cauchy-Schwarz inequality for operators with application*, Linear Algebra Appl., **223/224** (1995), 119–129.
12. R. Bhatia and F. Kittaneh, *On the singular values of a product of operators*, SIAM J. Matrix Anal. Appl., **11** (1990), 272–277.
13. R. Bhatia and K. R. Parthasarathy, *Positive definite functions and operator inequalities*, Bull. London Math. Soc., **32** (2000), 214–228.
14. M. Sh. Birman and M. Z. Solomyak, *Stieltjes double operator integrals*, Dokl. Akad. Nauk SSSR, **165** (1965), 1223-1226 (Russian); Soviet Math. Dokl., **6** (1965), 1567–1571.
15. M. Sh. Birman and M. Z. Solomyak, *Stieltjes double-integral operators*, Topics in Mathematical Physics, Vol. 1, M. Sh. Birman (ed.), Consultants Bureau, New York, 1967, pp. 25–54.
16. M. Sh. Birman and M. Z. Solomyak, *Stieltjes double-integral operators. II*, Topics in Mathematical Physics, Vol. 2, M. Sh. Birman (ed.), Consultants Bureau, New York, 1968, pp. 19–46.
17. M. Sh. Birman and M. Z. Solomyak, *Spectral Theory of Self-Adjoint Operators in Hilbert Space*, D. Reidel Publishing Company, Dordrecht, 1986.

18. V. I. Chilin, A. Krygin and F. A. Sukochev, *Uniform convexity and local uniform convexity of symmetric spaces of measurable operators*, Dokl. Akad. Nauk SSSR, **317** (1991), 555–558 (Russian); Soviet Math. Dokl., **43** (1991), 445–448.

19. G. Corach. H. Porta, and L. Recht, *An operator inequality*, Linear Algebra Appl., **142** (1990), 153–158.

20. G. Corach. H. Porta, and L. Recht, *A geometric interpretation of Segal's inequality* $\|e^{X+Y}\| \leq \|e^{X/2}e^Y e^{X/2}\|$, Proc. Amer. Math. Soc., **115** (1992), 229–231.

21. C. C. Cowen, K. E. Debro and P. D. Sepanski, *Geometry and the norms of Hadamard multipliers*, Linear Algebra Appl., **218** (1995), 239–249.

22. A. Van Daele, *A new approach to the Tomita-Takesaki theory of generalized Hilbert algebras*, J. Funct. Anal., **15** (1974), 378–393.

23. A. Van Daele and M. Rieffel, *A bounded operator approach to Tomita-Takesaki theory*, Pacific J. Math., **69** (1977), 187-221.

24. E. B. Davies, *Lipschitz continuity of functions of operators in the Schatten classes*, J. London Math. Soc. (2), **37** (1988), 148–157.

25. N. Dunford and J. Schwartz, *Linear Operators Part I: General Theory*, Wiley-Interscience, New York, 1988.

26. T. Fack and H. Kosaki, *Generalized s-numbers of τ-measurable operators*, Pacific J. Math., **123** (1986), 269–300.

27. Yu. B. Farforovskaya, *An estimate of the norm $\|f(A) - f(B)\|$ for self-adjoint operators A and B*, Zap. Nauch. Sem. LOMI, **56** (1976), 143–162.

28. J. Fujii, M. Fujii, T. Furuta and R. Nakamoto, *Norm inequalities equivalent to Heinz inequality*, Proc. Amer. Math. Soc., **118** (1993), 827–830.

29. I. C. Gohberg and M. G. Krein, *Introduction to the Theory of Linear Non-selfadjoint Operators*, Translations of Mathematical Monographs, Vol. 18, Amer. Math. Soc., Providence, 1969.

30. I. C. Gohberg and M. G. Krein, *Theory and Applications of Volterra Operators in Hilbert Spaces*, Translations of Mathematical Monographs, Vol. 24, Amer. Math. Soc., Providence, 1970.

31. U. Haagerup, *On Schur multipliers in $\mathcal{C}_1$*, unpublished hand-written notes (1980).

32. U. Haagerup, *Decomposition of completely bounded maps on operator algebras*, unpublished hand-written notes (1980).

33. F. Hansen and G. K. Pedersen, *Perturbation formulas for traces on C^*-algebras*, Publ. Res. Inst. Math. Sci., **31** (1995), 169–178.

34. G. Hardy, J. E. Littlewood and G. Pólya, *Inequalities* (Second edition), Cambridge Univ. Press, 1952.

35. M. Hasumi, *The extension property of complex Banach spaces*, Tôhoku Math. J., **10** (1958), 135–142.

36. E. Heinz, *Beiträge zur Störungstheorie der Spektralzerlegung*, Math. Ann., **123** (1951), 415–438.

37. F. Hiai, *Log-majorizations and norm inequalities for exponential operators*, Linear Operators, Banach Center Publications, Vol. 38, Polish Academy of Sciences, Warszawa, 1997, pp. 119–181.

38. F. Hiai and H. Kosaki, *Comparison of various means for operators*, J. Funct. Anal., **163** (1999), 300–323.

39. F. Hiai and H. Kosaki, *Means for matrices and comparison of their norms*, Indiana Univ. Math. J., **48** (1999), 899–936.

40. F. Hiai and H. Kosaki, *Operator means and their norms*, in "Operator Algebras and Applications" to appear in ASPM.

41. R. A. Horn, *Norm bounds for Hadamard products and the arithmetic-geometric mean inequality for unitarily invariant norms*, Linear Algebra Appl., **223/224** (1995), 355–361.

42. R. A. Horn and C. R. Johnson, *Topics in Matrix Analysis*, Cambridge Univ. Press, 1990.

43. R. A. Horn and X. Zhan, *Inequalities for C-S seminorms and Lieb functions*, Linear Algebra Appl., **291** (1999), 103-113.

44. T. Itoh and M. Nagisa, *Numerical radius norm for bounded module maps and Schur multipliers*, preprint (2002).

45. D. J. Jocić, *Norm inequalities for self-adjoint derivations*, J. Funct. Anal., **145** (1997), 24–34.

46. D. J. Jocić and F. Kittaneh, *Some perturbation inequalities for self-adjoint operators*, J. Operator Theory, **171** (1994), 3–10.

47. L. V. Kantorovich and G. P. Akilov, *Functional Analysis* (Second edition), Pergamon Press, Oxford, 1982.

48. T. Kato, *Perturbation Theory for Linear Operators* (Second edition), Springer-Verlag, Berlin-New York, 1976.

49. Y. Katznelson, *An Introduction to Harmonic Analysis* (Second corrected edition), Dover, New York, 1976.

50. F. Kittaneh, *A note on the arithmetic-geometric mean inequality for matrices*, Linear Algebra Appl., **171** (1992), 1–8.

51. F. Kittaneh, *On some operator inequalities*, Linear Algebra Appl., **208/209** (1994), 19–28.

52. L. S. Koplienko, *On the theory of the spectral shift function*, Topics in Mathematical Physics, Vol. 5, M. Sh. Birman (ed.), Consultants Bureau, New York, 1972, pp. 51–59.

53. H. Kosaki, *Unitarily invariant norms under which the map $A \to |A|$ is Lipschitz continuous*, Publ. Res. Inst. Math. Sci., **28** (1992), 299–313.

54. H. Kosaki, *Arithmetic-geometric mean and related inequalities for operators*, J. Funct. Anal., **156** (1998), 429–451.

55. H. Kosaki, in preparation.

56. S. G. Krein, Ju. I. Petunin, and E. M. Semenov, *Interpolation of Linear Operators*, Translations of Mathematical Monographs Vol. 54, Amer. Math. Soc., Providence, 1982.

57. F. Kubo and T. Ando, *Means of positive linear operators*, Math. Ann. 246 (1980), 205–224.

58. S. T. Kuroda, *On a theorem of Weyl-von Neumann*, Proc. Japan Acad., **34** (1958), 11–15.

59. S. Kwapień and A. Pełczyński, *The main triangular projection in matrix spaces and its applications*, Studia Math., **34** (1970), 43–68.

60. S. Lang, *Fundamentals of Differential Geometry*, Springer-Verlag, 1999.

61. J. Lindenstrauss and A. Pełczyński, *Absolutely summing operators in $\mathcal{L}_p$-spaces and their applications*, Studia Math., **29** (1968), 275–326.

62. A. W. Marshall and I. Olkin, *Inequalities: Theory of Majorization and Its Applications*, Academic Press, New York, 1979.

63. R. Mathias, *An Arithmetic-Geometric-Harmonic mean inequality involving Hadamard products*, Linear Algebra Appl., **184** (1993), 71–78.

64. A. McIntosh, *Heinz inequalities and perturbation of spectral families*, Macquarie Mathematical Reports, 79-0006, 1979.

65. F. Oberhettinger, *Tables of Fourier Transforms and Fourier Transforms of Distributions*, Springer-Verlag, Berlin, 1990.

66. V. Paulsen, *Completely Bounded Maps and Dilations*, Pitman Res. Notes in Math. Ser., Vol. 146, Longmann, Harlow, 1986.

67. G. K. Pedersen, *On the operator equation $HT + TH = K$*, Indiana Univ. Math. J., **25** (1976), 1029–1033.

68. G. K. Pedersen, *Operator differentiable functions*, Publ. Res. Inst. Math. Sci., **36** (2000), 139–157.

69. V. V. Peller, *Hankel operators and differentiability properties of functions of self-adjoint (unitary) operators*, LOMI Preprints E-1-84, USSR Academy of Sciences Steklov Mathematical Institute Leningrad Department, 1984.

70. V. V. Peller, *Hankel operators in the perturbation theory of unitary and self-adjoint operators*, Funct. Anal. Appl., **19** (1985), 111–123.

71. V. V. Peller, *Hankel Operators and Their Applications*, Springer-Verlag, New York-Berlin-Heidelberg, 2003.

72. A. Pietsch, *Operator Ideals*, North-Holland, Amsterdam-New York-Oxford, 1980.

73. W. Pusz and S. L. Woronowicz, *Functional calculus for sesquilinear forms and the purification map*, Rep. Math. Phys., **8** (1975), 159–170.

74. M. Reed and B. Simon, *Methods of Modern Mathematical Physics I: Functional Analysis*, Academic Press, New York, 1980.

75. H. L. Royden, *Real Analysis* (Second edition), Macmillan, New York, 1968.

76. H. H. Schaefer, *Banach Lattices and Positive Operators*, Springer-Verlag, Berlin-Heidelberg-New York, 1974

77. B. Simon, *Trace Ideals and Their Applications*, Cambridge Univ. Press, Cambridge, 1979.

78. M. Singh, J. S. Aujla, and H. L. Vesudeva, *Inequalities for Hadamard product and unitarily invariant norms of matrices*, Linear and Multilinear Algebra, **48** (2001), 247-262.

79. F. Smithies, *Integral Equations*, Cambridge Univ. Press, Cambridge, 1958.

80. F. C. Titchmarsh, *The Theory of Functions* (Second edition), Oxford Univ. Press, London, 1939.

81. F. G. Tricomi, *Integral Equations*, Interscience Publ., New York, 1957.

82. K. Yosida, *Lectures on Differential and Integral Equations*, Interscience Publ., New York, 1960.

83. X. Zhan, *Inequalities for unitarily invariant norms*, SIAM J. Matrix Anal. Appl., **20** (1998), 466-470.

84. X. Zhan, *Matrix Inequalities*, Lecture Notes in Math., Vol. **1790**, Springer-Verlag, 2002.

Index

Lecture Notes in Mathematics

For information about Vols. 1–1639
please contact your bookseller or Springer-Verlag

Vol. 1685: S. König, A. Zimmermann, Derived Equivalences for Group Rings. X, 146 pages. 1998.

Vol. 1686: J. Azéma, M. Émery, M. Ledoux, M. Yor (Eds.), Séminaire de Probabilités XXXII. VI, 440 pages. 1998.

Vol. 1687: F. Bornemann, Homogenization in Time of Singularly Perturbed Mechanical Systems. XII, 156 pages. 1998.

Vol. 1688: S. Assing, W. Schmidt, Continuous Strong Markov Processes in Dimension One. XII, 137 page. 1998.

Vol. 1689: W. Fulton, P. Pragacz, Schubert Varieties and Degeneracy Loci. XI, 148 pages. 1998.

Vol. 1690: M. T. Barlow, D. Nualart, Lectures on Probability Theory and Statistics. Editor: P. Bernard. VIII, 237 pages. 1998.

Vol. 1691: R. Bezrukavnikov, M. Finkelberg, V. Schechtman, Factorizable Sheaves and Quantum Groups. X, 282 pages. 1998.

Vol. 1692: T. M. W. Eyre, Quantum Stochastic Calculus and Representations of Lie Superalgebras. IX, 138 pages. 1998.

Vol. 1694: A. Braides, Approximation of Free-Discontinuity Problems. XI, 149 pages. 1998.

Vol. 1695: D. J. Hartfiel, Markov Set-Chains. VIII, 131 pages. 1998.

Vol. 1696: E. Bouscaren (Ed.): Model Theory and Algebraic Geometry. XV, 211 pages. 1998.

Vol. 1697: B. Cockburn, C. Johnson, C.-W. Shu, E. Tadmor, Advanced Numerical Approximation of Nonlinear Hyperbolic Equations. Cetraro, Italy, 1997. Editor: A. Quarteroni. VII, 390 pages. 1998.

Vol. 1698: M. Bhattacharjee, D. Macpherson, R. G. Möller, P. Neumann, Notes on Infinite Permutation Groups. XI, 202 pages. 1998.

Vol. 1699: A. Inoue, Tomita-Takesaki Theory in Algebras of Unbounded Operators. VIII, 241 pages. 1998.

Vol. 1700: W. A. Woyczyński, Burgers-KPZ Turbulence, XI, 318 pages. 1998.

Vol. 1701: Ti-Jun Xiao, J. Liang, The Cauchy Problem of Higher Order Abstract Differential Equations, XII, 302 pages. 1998.

Vol. 1702: J. Ma, J. Yong, Forward-Backward Stochastic Differential Equations and Their Applications. XIII, 270 pages. 1999.

Vol. 1703: R. M. Dudley, R. Norvaiša, Differentiability of Six Operators on Nonsmooth Functions and p-Variation. VIII, 272 pages. 1999.

Vol. 1704: H. Tamanoi, Elliptic Genera and Vertex Operator Super-Algebras. VI, 390 pages. 1999.

Vol. 1705: I. Nikolaev, E. Zhuzhoma, Flows in 2-dimensional Manifolds. XIX, 294 pages. 1999.

Vol. 1706: S. Yu. Pilyugin, Shadowing in Dynamical Systems. XVII, 271 pages. 1999.

Vol. 1707: R. Pytlak, Numerical Methods for Optimal Control Problems with State Constraints. XV, 215 pages. 1999.

Vol. 1708: K. Zuo, Representations of Fundamental Groups of Algebraic Varieties. VII, 139 pages. 1999.

Vol. 1709: J. Azéma, M. Émery, M. Ledoux, M. Yor (Eds), Séminaire de Probabilités XXXIII. VIII, 418 pages. 1999.

Vol. 1710: M. Koecher, The Minnesota Notes on Jordan Algebras and Their Applications. IX, 173 pages. 1999.

Vol. 1711: W. Ricker, Operator Algebras Generated by Commuting Projéctions: A Vector Measure Approach. XVII, 159 pages. 1999.

Vol. 1712: N. Schwartz, J. J. Madden, Semi-algebraic Function Rings and Reflectors of Partially Ordered Rings. XI, 279 pages. 1999.

Vol. 1713: F. Bethuel, G. Huisken, S. Müller, K. Steffen, Calculus of Variations and Geometric Evolution Problems. Cetraro, 1996. Editors: S. Hildebrandt, M. Struwe. VII, 293 pages. 1999.

Vol. 1714: O. Diekmann, R. Durrett, K. P. Hadeler, P. K. Maini, H. L. Smith, Mathematics Inspired by Biology. Martina Franca, 1997. Editors: V. Capasso, O. Diekmann. VII, 268 pages. 1999.

Vol. 1715: N. V. Krylov, M. Röckner, J. Zabczyk, Stochastic PDE's and Kolmogorov Equations in Infinite Dimensions. Cetraro, 1998. Editor: G. Da Prato. VIII, 239 pages. 1999.

Vol. 1716: J. Coates, R. Greenberg, K. A. Ribet, K. Rubin, Arithmetic Theory of Elliptic Curves. Cetraro, 1997. Editor: C. Viola. VIII, 260 pages. 1999.

Vol. 1717: J. Bertoin, F. Martinelli, Y. Peres, Lectures on Probability Theory and Statistics. Saint-Flour, 1997. Editor: P. Bernard. IX, 291 pages. 1999.

Vol. 1718: A. Eberle, Uniqueness and Non-Uniqueness of Semigroups Generated by Singular Diffusion Operators. VIII, 262 pages. 1999.

Vol. 1719: K. R. Meyer, Periodic Solutions of the N-Body Problem. IX, 144 pages. 1999.

Vol. 1720: D. Elworthy, Y. Le Jan, X-M. Li, On the Geometry of Diffusion Operators and Stochastic Flows. IV, 118 pages. 1999.

Vol. 1721: A. Iarrobino, V. Kanev, Power Sums, Gorenstein Algebras, and Determinantal Loci. XXVII, 345 pages. 1999.

Vol. 1722: R. McCutcheon, Elemental Methods in Ergodic Ramsey Theory. VI, 160 pages. 1999.

Vol. 1723: J. P. Croisille, C. Lebeau, Diffraction by an Immersed Elastic Wedge. VI, 134 pages. 1999.

Vol. 1724: V. N. Kolokoltsov, Semiclassical Analysis for Diffusions and Stochastic Processes. VIII, 347 pages. 2000.

Vol. 1725: D. A. Wolf-Gladrow, Lattice-Gas Cellular Automata and Lattice Boltzmann Models. IX, 308 pages. 2000.

Vol. 1726: V. Marić, Regular Variation and Differential Equations. X, 127 pages. 2000.

Vol. 1727: P. Kravanja M. Van Barel, Computing the Zeros of Analytic Functions. VII, 111 pages. 2000.

Vol. 1728: K. Gatermann Computer Algebra Methods for Equivariant Dynamical Systems. XV, 153 pages. 2000.

Vol. 1729: J. Azéma, M. Émery, M. Ledoux, M. Yor Séminaire de Probabilités XXXIV. VI, 431 pages. 2000.

Vol. 1730: S. Graf, H. Luschgy, Foundations of Quantization for Probability Distributions. X, 230 pages. 2000.

Vol. 1731: T. Hsu, Quilts: Central Extensions, Braid Actions, and Finite Groups. XII, 185 pages. 2000.

Vol. 1732: K. Keller, Invariant Factors, Julia Equivalences and the (Abstract) Mandelbrot Set. X, 206 pages. 2000.

Vol. 1733: K. Ritter, Average-Case Analysis of Numerical Problems. IX, 254 pages. 2000.

Vol. 1734: M. Espedal, A. Fasano, A. Mikelić, Filtration in Porous Media and Industrial Applications. Cetraro 1998. Editor: A. Fasano. 2000.

Vol. 1735: D. Yafaev, Scattering Theory: Some Old and New Problems. XVI, 169 pages. 2000.

Vol. 1736: B. O. Turesson, Nonlinear Potential Theory and Weighted Sobolev Spaces. XIV, 173 pages. 2000.

Vol. 1788: A. Vasil'ev, Moduli of Families of Curves for Conformal and Quasiconformal Mappings.IX, 211 pages. 2002.

Vol. 1789: Y. Sommerhäuser, Yetter-Drinfel'd Hopf algebras over groups of prime order. V, 157 pages. 2002.

Vol. 1790: X. Zhan, Matrix Inequalities. VII, 116 pages. 2002.

Vol. 1791: M. Knebusch, D. Zhang, Manis Valuations and Prüfer Extensions I: A new Chapter in Commutative Algebra. VI, 267 pages. 2002.

Vol. 1792: D. D. Ang, R. Gorenflo, V. K. Le, D. D. Trong, Moment Theory and Some Inverse Problems in Potential Theory and Heat Conduction. VIII, 183 pages. 2002.

Vol. 1793: J. Cortés Monforte, Geometric, Control and Numerical Aspects of Nonholonomic Systems. XV, 219 pages. 2002.

Vol. 1794: N. Pytheas Fogg, Substitution in Dynamics, Arithmetics and Combinatorics. Editors: V. Berthé, S. Ferenczi, C. Mauduit, A. Siegel. XVII, 402 pages. 2002.

Vol. 1795: H. Li, Filtered-Graded Transfer in Using Noncommutative Gröbner Bases. IX, 197 pages. 2002.

Vol. 1796: J.M. Melenk, hp-Finite Element Methods for Singular Perturbations. XIV, 318 pages. 2002.

Vol. 1797: B. Schmidt, Characters and Cyclotomic Fields in Finite Geometry. VIII, 100 pages. 2002.

Vol. 1798: W.M. Oliva, Geometric Mechanics. XI, 270 pages. 2002.

Vol. 1799: H. Pajot, Analytic Capacity, Rectifiability, Menger Curvature and the Cauchy Integral. XII,119 pages. 2002.

Vol. 1800: O. Gabber, L. Ramero, Almost Ring Theory. VI, 307 pages. 2003.

Vol. 1801: J. Azéma, M. Émery, M. Ledoux, M. Yor, Séminaire de Probabilités XXXVI. VIII, 499 pages. 2003.

Vol. 1802: V. Capasso, E. Merzbach, B.G. Ivanoff, M. Dozzi, R. Dalang, T. Mountford, Topics in Spatial Stochastic Processes. Martina Franca, Italy 2001. Editor: E. Merzbach. VIII, 253 pages. 2003.

Vol. 1803: G. Dolzmann, Variational Methods for Crystalline Microstructure - Analysis and Computation. VIII, 212 pages. 2003.

Vol. 1804: I. Cherednik, Ya. Markov, R. Howe, G. Lusztig, Iwahori-Hecke Algebras and their Representation Theory. Martina Franca, Italy 1999. Editors: V. Baldoni, D. Barbasch. X, 103 pages. 2003.

Vol. 1805: F. Cao, Geometric Curve Evolution and Image Processing. X, 187 pages. 2003.

Vol. 1806: H. Broer, I. Hoveijn. G. Lunther, G. Vegter, Bifurcations in Hamiltonian Systems. Computing Singularities by Gröbner Bases. XIV, 169 pages. 2003.

Vol. 1807: V. D. Milman, G. Schechtman, Geometric Aspects of Functional Analysis. Israel Seminar 2000-2002. VIII, 429 pages. 2003.

Vol. 1808: W. Schindler, Measures with Symmetry Properties.IX, 167 pages. 2003.

Vol. 1809: O. Steinbach, Stability Estimates for Hybrid Coupled Domain Decomposition Methods. VI, 120 pages. 2003.

Vol. 1810: J. Wengenroth, Derived Functors in Functional Analysis. VIII, 134 pages. 2003.

Vol. 1811: J. Stevens, Deformations of Singularities. VII, 157 pages. 2003.

Vol. 1812: L. Ambrosio, K. Deckelnick, G. Dziuk, M. Mimura, V. A. Solonnikov, H. M. Soner, Mathematical Aspects of Evolving Interfaces. Madeira, Funchal, Portugal 2000. Editors: P. Colli, J. F. Rodrigues. X, 237 pages. 2003.

Vol. 1813: L. Ambrosio, L. A. Caffarelli, Y. Brenier, G. Buttazzo, C. Villani, Optimal Transportation and its Applications. Martina Franca, Italy 2001. Editors: L. A. Caffarelli, S. Salsa. X, 164 pages. 2003.

Vol. 1814: P. Bank, F. Baudoin, H. Föllmer, L.C.G. Rogers, M. Soner, N. Touzi, Paris-Princeton Lectures on Mathematical Finance. X,172 pages. 2003.

Vol. 1815: A. M. Vershik (Ed.), Asymptotic Combinatorics with Applications to Mathematical Physics. St. Petersburg, Russia 2001. IX, 246 pages. 2003.

Vol. 1816: S. Albeverio, W. Schachermayer, M. Talagrand, Lectures on Probability Theory and Statistics. Ecole d'Eté de Probabilités de Saint-Flour XXX-2000. Editor: P. Bernard. VIII, 296 pages. 2003.

Vol. 1817: E. Koelink (Ed.), Orthogonal Polynomials and Special Functions. Leuven 2002. X, 249 pages. 2003.

Vol. 1818: M. Bildhauer, Convex Variational Problems with Linear, nearly Linear and/or Anisotropic Growth Conditions. X, 217 pages. 2003.

Vol. 1819: D. Masser, Yu. V. Nesterenko, H. P. Schlickewei, W. M. Schmidt, M. Waldschmidt, Diophantine Approximation. Cetraro, Italy 2000. Editors: F. Amoroso, U. Zannier. XI,353 pages. 2003.

Vol. 1820: F. Hiai, H. Kosaki, Means of Hilbert Space Operators. VIII, 148 pages. 2003.

Vol. 1821: S. Teufel, Adiabatic Perturbation Theory in Quantum Dynamics.VI, 236 pages. 2003.

Vol. 1822: S.-N. Chow, R. Conti, R. Johnson, J. Mallet-Paret, R. Nussbaum, Dynamical Systems. Cetraro, Italy 2000. Editors: J. W. Macki, P. Zecca. XII, 345 pages. 2003.

Vol. 1823: A. M. Anile, W. Allegretto, C. Ringhofer, Mathematical Problems in Semiconductor Physics. Cetraro, Italy 1998. Editor: A. M. Anile. X, 135 pages. 2003.

Vol. 1824: J. A. Navarro González, J. B. Sancho de Salas, C^∞ - Differentiable Spaces. XIII, 188 pages. 2003.

Vol. 1825: J. H. Bramble, A. Cohen, W. Dahmen, Multiscale Problems and Methods in Numerical Simulations, Martina Franca, Italy 2001. Editor: C. Canuto. XII, 155 pages. 2003.

Recent Reprints and New Editions

Vol. 1200: V. D. Milman, G. Schechtman, Asymptotic Theory of Finite Dimensional Normed Spaces. 1986. – Corrected Second Printing. X, 156 pages. 2001.

Vol. 1618: G. Pisier, Similarity Problems and Completely Bounded Maps. 1995 – Second, Expanded Edition VII, 198 pages. 2001.

Vol. 1629: J. D. Moore, Lectures on Seiberg-Witten Invariants. 1997 – Second Edition. VIII, 121 pages. 2001.

Vol. 1638: P. Vanhaecke, Integrable Systems in the realm of Algebraic Geometry. 1996 – Second Edition. X, 256 pages. 2001.

Vol. 1702: J. Ma, J. Yong, Forward-Backward Stochastic Differential Equations and Their Applications. 1999. – Corrected Second Printing. XIII, 270 pages. 2000.